Hepatocyte Transplantation

METHODS IN MOLECULAR BIOLOGY™

John M. Walker, SERIES EDITOR

502. Bacteriophages: *Methods and Protocols, Volume 2: Molecular and Applied Aspects,* edited by *Martha R. J. Clokie and Andrew M. Kropinski 2009*
501. Bacteriophages: *Methods and Protocols, Volume 1: Isolation,* Characterization, and Interactions, edited by *Martha R. J. Clokie and Andrew M. Kropinski 2009*
496. DNA and RNA Profiling in Human Blood: *Methods and Protocols,* edited by *Peter Bugert, 2009*
493. Auditory and Vestibular Research: *Methods and Protocols,* edited by *Bernd Sokolowski, 2009*
490. Protein Structures, Stability, and Interactions, edited by *John W. Schriver, 2009*
489. Dynamic Brain Imaging: *Methods and Protocols,* edited by *Fahmeed Hyder, 2009*
485. HIV Protocols: *Methods and Protocols,* edited by *Vinayaka R. Prasad and Ganjam V. Kalpana, 2009*
484. Functional Proteomics: *Methods and Protocols,* edited by *Julie D. Thompson, Christine Schaeffer-Reiss, and Marius Ueffing, 2008*
483. Recombinant Proteins From Plants: *Methods and Protocols,* edited by *Lóic Faye and Veronique Gomord, 2008*
482. Stem Cells in Regenerative Medicine: *Methods and Protocols,* edited by *Julie Audet and William L. Stanford, 2008*
481. Hepatocyte Transplantation: *Methods and Protocols,* edited by *Anil Dhawan and Robin D. Hughes, 2009*
480. Macromolecular Drug Delivery: *Methods and Protocols,* edited by *Mattias Belting, 2008*
479. Plant Signal Transduction: *Methods and Protocols,* edited by *Thomas Pfannschmidt, 2008*
478. Transgenic Wheat, Barley and Oats: *Production and Characterization Protocols,* edited by *Huw D. Jones and Peter R. Shewry, 2008*
477. Advanced Protocols in Oxidative Stress I, edited by *Donald Armstrong, 2008*
476. Redox-Mediated Signal Transduction: *Methods and Protocols,* edited by *John T. Hancock, 2008*
475. Cell Fusion: *Overviews and Methods,* edited by *Elizabeth H. Chen, 2008*
474. Nanostructure Design: *Methods and Protocols,* edited by *Ehud Gazit and Ruth Nussinov, 2008*
473. Clinical Epidemiology: *Practice and Methods,* edited by *Patrick Parfrey and Brendon Barrett, 2008*
472. Cancer Epidemiology, Volume 2: *Modifiable Factors,* edited by *Mukesh Verma, 2008*
471. Cancer Epidemiology, Volume 1: *Host Susceptibility Factors,* edited by *Mukesh Verma, 2008*
470. Host-Pathogen Interactions: *Methods and Protocols,* edited by *Steffen Rupp and Kai Sohn, 2008*
469. Wnt Signaling, Volume 2: *Pathway Models,* edited by *Elizabeth Vincan, 2008*
468. Wnt Signaling, Volume 1: *Pathway Methods and Mammalian Models,* edited by *Elizabeth Vincan, 2008*
467. Angiogenesis Protocols: *Second Edition,* edited by *Stewart Martin and Cliff Murray, 2008*
466. Kidney Research: *Experimental Protocols,* edited by *Tim D. Hewitson and Gavin J. Becker, 2008*
465. Mycobacteria, Second Edition, edited by *Tanya Parish and Amanda Claire Brown, 2008*
464. The Nucleus, Volume 2: *Physical Properties and Imaging Methods,* edited by *Ronald Hancock, 2008*
463. The Nucleus, Volume 1: *Nuclei and Subnuclear Components,* edited by *Ronald Hancock, 2008*
462. Lipid Signaling Protocols, edited by *Banafshe Larijani, Rudiger Woscholski, and Colin A. Rosser, 2008*
461. Molecular Embryology: *Methods and Protocols, Second Edition,* edited by *Paul Sharpe and Ivor Mason, 2008*
460. Essential Concepts in Toxicogenomics, edited by *Donna L. Mendrick and William B. Mattes, 2008*
459. Prion Protein Protocols, edited by *Andrew F. Hill, 2008*
458. Artificial Neural Networks: *Methods and Applications,* edited by *David S. Livingstone, 2008*
457. Membrane Trafficking, edited by *Ales Vancura, 2008*
456. Adipose Tissue Protocols, Second Edition, edited by *Kaiping Yang, 2008*
455. Osteoporosis, edited by *Jennifer J. Westendorf, 2008*
454. SARS- and Other Coronaviruses: *Laboratory Protocols,* edited by *Dave Cavanagh, 2008*
453. Bioinformatics, Volume 2: *Structure, Function, and Applications,* edited by *Jonathan M. Keith, 2008*
452. Bioinformatics, Volume 1: *Data, Sequence Analysis, and Evolution,* edited by *Jonathan M. Keith, 2008*
451. Plant Virology Protocols: *From Viral Sequence to Protein Function,* edited by *Gary Foster, Elisabeth Johansen, Yiguo Hong, and Peter Nagy, 2008*
450. Germline Stem Cells, edited by *Steven X. Hou and Shree Ram Singh, 2008*
449. Mesenchymal Stem Cells: *Methods and Protocols,* edited by *Darwin J. Prockop, Douglas G. Phinney, and Bruce A. Brunnell, 2008*
448. Pharmacogenomics in Drug Discovery and Development, edited by *Qing Yan, 2008*
447. Alcohol: *Methods and Protocols,* edited by *Laura E. Nagy, 2008*
446. Post-translational Modifications of Proteins: *Tools for Functional Proteomics, Second Edition,* edited by *Christoph Kannicht, 2008*
445. Autophagosome and Phagosome, edited by *Vojo Deretic, 2008*
444. Prenatal Diagnosis, edited by *Sinhue Hahn and Laird G. Jackson, 2008*
443. Molecular Modeling of Proteins, edited by *Andreas Kukol, 2008*

METHODS IN MOLECULAR BIOLOGY™

Hepatocyte Transplantation

Methods and Protocols

Edited by

Anil Dhawan

King's College Hospital, London, UK

Robin D. Hughes

King's College London, School of Medicine
London, UK

Humana Press

Editors
Anil Dhawan
King's College Hospital
London, UK
anil.dhawan@kcl.ac.uk

Robin D. Hughes
King's College London
School of Medicine
London, UK
robin.hughes@kcl.ac.uk

Series Editor
John M. Walker
University of Hertfordshire
Hatfield, Herts.
UK

ISBN: 978-1-58829-883-6
e-ISBN: 978-1-59745-201-4
ISSN: 1064-3745
e-ISSN: 1940-6029
DOI 10.1007/978-1-59745-201-4

Library of Congress Control Number: 2008939645

Cover illustration: Figure 1 from chapter 15

Printed on acid-free paper

9 8 7 6 5 4 3 2 1

springer.com

Preface

Cellular therapy using human hepatocytes is being evaluated worldwide as an alternative to organ transplantation in patients with liver-based metabolic disease and acute liver failure. The basis for clinical use has come from the demonstration of efficacy in animal models of acute and chronic liver disease.

Protocols have been developed for the isolation of hepatocytes from liver tissue under GMP conditions and also for improved methods of cryopreservation, so hepatocytes can be stored for later clinical use. Assays are used to assess the quality and function of the hepatocytes prior to transplantation. There are clinical protocols for administration of cells directly into the patient's liver.

The engraftment of donor cells in the recipient liver can be detected by DNA techniques or functional proteins in the case of genetic liver disorders. In vivo methods are needed to track the fate of hepatocytes after transplantation.

Due to the shortage of donor organs, the future of hepatocyte transplantation will be alternative sources of liver cells such as foetal hepatoblasts or stem cell-derived hepatocytes. Methods for culture and in vitro proliferation of stem cells will be important for their application.

It is hoped that this volume from the experts in the field provides the reader with the practical protocols to enable them to perform and investigate hepatocyte transplantation. Needless to say this is a rapidly developing field, and new and improved techniques are being developed all the time.

Anil Dhawan & Robin D. Hughes

Acknowledgements

To my wife Anita and boys Atin and Ashish for their understanding, love and support that they have provided throughout my career.

Sincere thanks to all the contributors.

Particular thanks to Professor Nigel Heaton, Mr Mohamed Rela, Liver Transplant Coordinators, and Dr Ragai Mitry for helping establish the hepatocyte transplantation programme at King's College Hospital.

Anil Dhawan

Contents

Contributors

MALCOLM R. ALISON • *Centre for Diabetes and Metabolic Medicine, ICMS, Bart's and The London School of Medicine, London, UK*

JOSÉ V. CASTELL • *Unit of Experimental Hepatology, University Hospital "La Fe", Valencia, Spain*

IBRAHIM DAGHER • *Inserm U 804; University Paris-Sud, Hôpital de Bicêtre, Kremlin-Bicêtre, and Service de Chirurgie Générale, Hôpital Béclère, Clamart, France*

JULIO DAVILA • *Pfizer, Inc., St. Louis Mo, USA*

ANIL DHAWAN • *Paediatric Liver Centre, King's College Hospital, Denmark Hill, London, UK*

EBTISAM EL FILALI • *AMC Liver center, Amsterdam, The Netherlands*

KEN DORKO • *Departments of Pathology and Surgery and McGowan Institute for Regenerative Medicine, University of Pittsburgh, USA*

EWA C.S. ELLIS • *Departments of Pathology and Surgery and McGowan Institute for Regenerative Medicine, University of Pittsburgh, USA*

ROBERT A. FISHER • *Department of Surgery, Transplantation Division, Virginia Commonwealth University, Medical College of Virginia Hospitals, Richmond, Virginia, USA*

DOMINIQUE FRANCO • *Inserm U 804; University Paris-Sud, Hôpital de Bicêtre, Kremlin-Bicêtre, and Service de Chirurgie Générale, Hôpital Béclère, Clamart, France*

AMANDA M. GIDDINGS • *BIBRA International, Carshalton, Surrey and Centre for Toxicology, Faculty of Health and Medical Sciences, University of Surrey, Guildford, UK*

MARÍA JOSÉ GÓMEZ-LECHÓN • *Unit of Experimental Hepatology, University Hospital "La Fe", Valencia, Spain*

ROBERTO GRAMIGNOLI • *Departments of Pathology and Surgery and McGowan Institute for Regenerative Medicine, University of Pittsburgh, USA*

MARIE-THÉRÈSE GROYER-PICARD • *Inserm U 804; University Paris-Sud, Hôpital de Bicêtre, Kremlin-Bicêtre, France*

SANJEEV GUPTA • *Marion Bessin Liver Research Center, Diabetes Center, Cancer Research Center, Departments of Medicine and Pathology, and Institute for Clinical and Translational Research, Albert Einstein College of Medicine, New York, USA*

ROBIN D. HUGHES • *Institute of Liver Studies, King's College London School of Medicine, London, UK*

ROSEMARY JEFFERY • *Histopathology Unit, Cancer Research UK, London Research Institute, London, UK*

RUTH JOPLIN • *Liver Research Laboratories, Institute of Biomedical Research, University of Birmingham Medical School, Birmingham, UK*

Stivelia Kachilele • *Liver Research Laboratories, Institute of Biomedical Research, University of Birmingham Medical School, Birmingham, UK*
Brian G. Lake • *BIBRA International, Carshalton, Surrey, and Centre for Toxicology, Faculty of Health and Medical Sciences, University of Surrey, Guildford, UK*
Sharon C. Lehec • *Institute of Liver Studies, King's College Hospital, London, UK*
Prashant Mani • *Department of Biochemistry, Delhi University South Campus, New Delhi, India*
Fabio Marongiu • *Departments of Pathology and Surgery and McGowan Institute for Regenerative Medicine, University of Pittsburgh, USA*
Valeria R. Mas • *Department of Surgery, Transplantation Division and Department of Pathology, Division of Molecular Diagnostics, Virginia Commonwealth University, Medical College of Virginia Hospitals, Richmond, Virginia, USA*
Thomas J. Meade • *Departments of Chemistry, Biochemistry, Molecular and Cell Biology, Neurobiology and Physiology, Northwestern University, Evanston, USA*
Toshio Miki • *Departments of Pathology and Surgery and McGowan Institute for Regenerative Medicine, University of Pittsburgh, USA*
Stephen L Minger • *Stem Cell Biology Laboratory, Wolfson Centre for Age-Related Diseases Kings College London, London, UK*
Keitaro Mitamura • *Departments of Pathology and Surgery and McGowan Institute for Regenerative Medicine, University of Pittsburgh, USA*
Ragai R. Mitry • *Institute of Liver Studies, King's College Hospital, London, UK*
Michel Modo • *Centre for the Cellular Basis of Behaviour, Institute of Psychiatry, King's College London, UK*
Mustapha Najimi • *Université Catholique de Louvain, Laboratory of Pediatric Hepatology & Cell Therapy, Brussels, Belgium*
Ronald Oude Elferink • *AMC Liver Center, Amsterdam, The Netherlands*
Richard Poulsom • *Histopathology Unit, Cancer Research UK, London Research Institute, London, UK*
Roger J. Price • *BIBRA International, Carshalton, Surrey and Centre for Toxicology, School of Biomedical and Molecular Sciences, University of Surrey, Guildford, UK*
Juliana Puppi • *Institute of Liver Studies, King's College London School of Medicine London, UK*
Aarati Ranade • *Departments of Pathology and Surgery and McGowan Institute for Regenerative Medicine, University of Pittsburgh, USA*
Namita Roy-Chowdhury • *Departments of Medicine and Molecular Genetics, and the Marion Bessin Liver Research Center, Albert Einstein College of Medicine, New York, USA*
Jayanta Roy-Chowdhury • *Departments of Medicine and Molecular Genetics, and the Marion Bessin Liver Research Center, Albert Einstein College of Medicine, New York, USA*
Niloufar Safinia • *Stem Cell Biology Laboratory, Wolfson Centre for Age-Related Diseases Kings College London, London, UK*
Debi P. Sarkar • *Department of Biochemistry, Delhi University South Campus, New Delhi, India*
Jurgen Seppen • *AMC Liver Center, Amsterdam, The Netherlands*

CHETAN C. SHATAPATHY • *Stem Cell Biology Laboratory, Wolfson Centre for Age-Related Diseases, King's College London, London, UK*

FRANÇOISE SMETS • *Université Catholique de Louvain, Laboratory of Pediatric Hepatology & Cell Therapy, Brussels, Belgium*

ETIENNE SOKAL • *Université Catholique de Louvain, Laboratory of Pediatric Hepatology & Cell Therapy, Brussels, Belgium*

STEPHEN C. STROM • *Departments of Pathology and Surgery and McGowan Institute for Regenerative Medicine, University of Pittsburgh, USA*

CLAIRE TERRY • *Institute of Liver Studies, King's College London School of Medicine London, UK*

DAVID G. WALTERS • *BIBRA International, Carshalton, Surrey, and Centre for Toxicology, Faculty of Health and Medical Sciences, University of Surrey, Guildford, UK*

XIA WANG • *Departments of Medicine and Molecular Genetics, and the Marion Bessin Liver Research Center, Albert Einstein College of Medicine, New York*

ANNE WEBER • *Inserm U 804; University Paris-Sud, Hôpital de Bicêtre, Kremlin-Bicêtre, France*

YAO-MING WU • *Department of Surgery, National Taiwan University Hospital, Taipei, Taiwan*

YUE WU • *Stem Cell Biology Laboratory, Wolfson Centre for Age-Related Diseases, King's College London, London, UK*

Color Plates

Color Plate 1: Apoptotic nuclei and bodies observed in mouse primary hepatocyte cultures after staurosporine treatment (white arrows). Freshly isolated mouse hepatocytes were plated for 24 h on a collagen type I-coated coverslips in well plates and treated for 4 h with 1 μM staurosporine. Cells were thereafter fixed with 4% of formaldehyde for 20 min at room temperature, stained with DAPI for 30 min and analyzed using a fluorescence microscopy. (*see* discussion on p. 63)

Color Plate 2: Condensation of chromatin at the periphery of the nucleus in apoptotic mouse hepatocytes (black arrows). (A) Primary mouse hepatocytes were plated for 24 h in a coated collagen type I well plates and treated for 4 h with 1 μM staurosporine. Cells were thereafter fixed with 4% formaldehyde for 20 min at room temperature and stained with HE for 10 min. (B) slice of mouse liver prefixed with formaldehyde, paraffin-embedded and HE-stained. (*see* discussion on p. 65)

Color Plate 3: Transplantation of autologous hepatocytes into *Macaca mulatta* after retroviral-mediated gene marking. (**A**) Protocol for simian hepatocyte isolation, retroviral transduction and transplantation. Hepatocyte transduction with HIV-1-derived lentivirus vectors avoids the culture steps. They are transduced in suspension and transplanted. (**B**) Hepatocytes are transplanted via the infusion chamber. (**C**) Freshly isolated simian hepatocytes at confluency after 3 days of culture. (**D**) Transduced hepatocytes in culture expressing the β-galactosidase. (**E**) Thawed hepatocytes after 3 days of culture. (*see* discussion on p. 90)

Color Plate 4: **Liver preconditioning using monocrotaline (MCT) for improving cell engraftment in DPPIV– rats.** Transplanted F344 rat hepatocytes are shown in the recipient liver 4 and 7 days after cell transplantation. Panel **a** shows 1–3 transplanted hepatocytes with histochemically visualized DPPIV activity (red color, arrows) in periportal areas (Pa). By contrast, in MCT-treated rats (**b**) several-fold more transplanted cells are present. Original magnification, ×200; hematoxylin counterstain. Modified from Joseph B, et al. (20). (*see* discussion on p. 111)

Color Plate 5: **Analysis of the kinetics of liver repopulation in DPPIV– rats preconditioned with retrorsine and partial hepatectomy**. Foci of transplanted cells with DPPIV activity (red color) are seen 2 (**a**), 3 (**b**), and 4 weeks (**c**) after cell transplantation. Morphometric analysis of liver repopulation in panel **d** indicates linear increase in liver repopulation during this period. Original magnification, (**a**–**c**), ×40; hematoxylin counterstain. Modified from Wu Y-M et al. 18. (*see* discussion on p. 112)

Color Plate 6: **Effect of immunosuppressive drugs, Rapamycin (Rapa) and Tacrolimus (Tacro), on liver repopulation in DPPIV– rats preconditioned with retrorsine and partial hepatectomy**. Animals were treated with drugs subsequent to the completion of cell engraftment. Rapa- but not Tacro-suppressed transplanted cell proliferation as shown by DPPIV histochemistry and morphometric analysis of either the extent of liver repopulation (**e**) or individual transplanted cell foci (**f**). Original magnification (**a**–**d**), ×100; hematoxylin counterstain. Modified from Wu Y-M et al. (18). (*see* discussion on p. 114)

Color Plate 7: **Transfection by Amaxa Nucleofection:** Expression of GFP in primary mouse hepatocytes (isolated from C57BL/6 mice) nucleofected using an Amaxa mouse hepatocyte Nucleofector kit with a plasmid encoding maxGFP. Twenty-four hours after nucleofection, cells were analyzed by bright field (**A**) and fluorescence microscopy (**B**). The merged image is shown in panel (**C**). (*see* discussion on p. 124)

Color Plate 8: **Transfection using liposomes containing F protein of the Sendai virus:** Expression of LacZ in cells transfected with DNA-loaded F-virosomes as described in the text. After incubation for 24 h, cells were fixed with ethanol, stained for β-galactosidase and photographed. (magnification, ×20, Nikon, Japan). Hepa1 cells (**A**), HEK293 cells (**B**). Note, only asialoglycoprotein-expressed cells are transduced by this method. Structure of histidine lipid used to enhance F-virosome-mediated gene transfer (**C**). (*see* discussion on p. 127)

Color Plate 9: **Transduction of primary rat hepatocytes using a Lentiviral vector:** Isolated Gunn rat hepatocytes were transduced with Lentivirus pAlb-UGT1A1 at an MOI of 10 and immunostained with WP1, monoclonal primary antibody against UGT1A1, followed by anti mouse Alkaline Phosphatase substrate kit III as described in the text and control hepatocytes (**A**) and experimental hepatocytes (**B**) were photographed. (*see* discussion on p. 132)

Color Plate 10: **Lentiviral vector-mediated transduction of primary mouse hepatocytes, enhanced by Magnetofection®:** Isolated mouse primary hepatocytes were transduced with Lentivirus pAlb-LacZ at an MOI of 5 with or without Magnetofection® as described in the text, and were stained 48 h later for bacterial β-galactosidase activity (blue reaction products). (**A**) Untransfected control; (**B**) Lentiviral transduction without Magnetofection®; (**C**) Lentiviral transduction enhanced by Magnetofection®. (*see* discussion on p. 133)

Color Plate 11: Revealing that bone marrow cells (BMCs) have differentiated into non-haematopoietic cells can be achieved by transplanting lethally irradiated animals with new BMCs that can be tracked whatever their subsequent fate. This would include male BMCs to a female recipient, or GFP- or LacZ-positive BMCs to wild-type recipients. The male chromosome can be detected by in situ hybridisation, GFP by immunohistochemistry and β-galactosidase by X-gal histochemistry. (*see* discussion on p. 141)

Color Plate 12: Fluorescent and confocal microscopy. (**A**) Male cells (arrows) in male bone marrow-transplanted female mouse liver (green FITC dot). These cells are CK18 immunoreactive (red cytoplasm), suggestive of hepatocyte differentiation. (**B**) Human cell (green FITC, spotty nucleus, arrowed) in mouse liver (pink CY3 spots) after injection of human $CD133^+$ cells into a NOD-SCID mouse. (**C**) BCR/ABL probe on human liver in a case of CML showing normal ploidy, with two copies of chromosome 9 (red signals) and two copies of chromosome 22 (green signals) in some cells (asterisks), but multiple copies (polyploidy) in another cell (arrow). (**D**) BCR/ABL fusion signal (green and red overlap producing orange, arrowed) seen in cell tentatively identified as a hepatocyte in a case of CML. There is one native chromosome 9 (red), one native chromosome 22 (green) and one small red signal (*ASS* gene). (**E**) Confocal images demonstrating liver polyploidy in a female mouse transplanted with male bone marrow, with multiple X chromosomes (green signals) showing that a Y chromosome (red signal, black arrow) is outside the nuclear membrane (view **E**), while a smaller nucleus (white arrow) has both X and Y chromosomes contained within it. (*see* discussion on p. 142)

Color Plate 13: Liver fibrosis in a mouse as viewed by bright field microscopy. (**A**) Demonstration of Y chromosome-positive cells (brown nuclear dots) in a female mouse liver after a male bone marrow transplant. (**B**) Demonstration of mRNA for pro(α1)I (black autoradiographic grains) in the same liver using a ^{3}H-labelled antisense riboprobe. (**C**) Demonstration of Y chromosome detection (brown dot, arrow) and IHC for α-SMA expression (red staining) – a marker of myofibroblast differentiation. (**D**) Demonstration of the expression of mRNA for pro(α1)I, the Y chromosome and α-SMA in the same liver. One Y chromosome-positive cell is expressing neither α-SMA nor mRNA for pro(α1)I, but another cell (asterisk) is expressing all three markers. Note the reduced grain density when techniques are combined in comparison to when ISH for the mRNA is performed alone. (**E** and **F**) Examples of ISH for pro(α1)I mRNA expression and immunoreactivity for α-SMA in the same section. (*see* discussion on p. 147)

Color Plate 14: The appearance of a Percoll gradient following centrifugation at 800×*g* for 30 min is shown. Layers 2 and 3 contain biliary epithelial cells (approximately 10%) and are harvested for further purification of immature and mBEC populations by immunomagnetic separation. The supernatant and fractions 1 and 4–6 are discarded. (*see* discussion on p. 197)

Color Plate 15: Visualisation of the MRI contrast agent. (**A**) Adult human hepatocytes being labelled with the bimodal Iron Oxide Green Oregon (IOGO) contrast agent (in green). Note that some cells (cell nuclei in blue) are not labelled. It is noteworthy that the contrast agent seems strongly associated with the cell nuclei and does not fill the cytoplasm. It is likely that mainly phagocytic Kupffer cells incorporated this agent, whereas unlabelled cells represent a small fraction of undifferentiated hepatocytes. (**B**) In contrast, the Gadolinium Rhodamine Dextran (GRID) bimodal agent (*in red*) clearly labels the cytoplasm of cells that have the appearance of immature hepatocytes and is incorporated into all types of cells. (*see* discussion on p. 212)

Chapter 1

Human Hepatocyte Transplantation Overview

Juliana Puppi and Anil Dhawan

Abstract

The interest in hepatocyte transplantation has been growing continuously in recent years and this treatment may represent an alternative clinical approach for patients with acute liver failure and liver-based metabolic disorders. This chapter presents an overview of liver cell transplantation, from the basic research to human experience. It summarizes the pre-clinical studies and present status of clinical hepatocyte transplantation and identifies some possible areas of future research in this area.

Key words: Hepatocyte transplantation, collagenase, cryopreservation, sources of liver tissues, GMP laboratory, clinical experience, future use

1. Introduction

Orthotopic liver transplantation (OLT) is the accepted method of treatment for end-stage liver disease and liver-based metabolic disorders. The improvements in patient and graft survival have mainly resulted from the developments in immunosuppressive drug therapy. Advances in surgical techniques now allow the use of auxiliary liver transplantation in the management of patients with acute liver failure (ALF) and certain liver-based metabolic defects such as Crigler–Najjar (CN) syndrome type I, urea cycle defects and familial hypercholesterolaemia. The success of auxiliary liver transplantation in humans *(1)* has supported the observation in animal experiments that relatively small amounts of liver tissue can provide sufficient function to correct the underlying metabolic defects. This has further increased the interest in using human hepatocytes for cell transplantation in the management of liver-based metabolic conditions and ALF.

Anil Dhawan, Robin D. Hughes (eds.), *Hepatocyte Transplantation, vol. 481*

DOI 10.1007/978-1-59745-201-4_1 Springerprotocols.com

There are a number of potential advantages of hepatocyte transplantation if the technique can be proved successful. It is less expensive and less invasive than OLT. It avoids the risks and undertaking of major surgery once liver cells can be transplanted after radiologic or surgical placement of a portal catheter. Unlike whole organs, hepatocytes can be cryopreserved and stored in cell banks, offering the advantage of immediate availability in emergencies. The transplanted cells functionally replace the hepatocytes of the diseased organ and restore its metabolic capacity either for a period of bridging to whole-organ transplantation or by engraftment and long-term function. Moreover, in hepatocyte transplantation, the recipient liver remains intact and subsequent liver-directed gene therapy would be still feasible when this becomes a clinical reality. With this there is the possibility of better utilization of donor organs, which remain in short supply, particularly if methods can be developed to isolate good-quality hepatocytes from marginal donor livers, currently rejected for clinical transplantation. Hepatocyte transplantation has been used as a treatment for ALF *(2–4)* and metabolic liver diseases such as CN syndrome type I *(5, 6)*, glycogen storage disease type 1a *(7)* and urea cycle defects *(8, 9)* for long-term correction of the underlying metabolic deficiency, with variable outcome.

2. Methods for Isolation of Human Hepatocytes

2.1. Sources of Liver Tissue

The major obstacle of liver cell therapy is the limited supply of donor liver tissue for hepatocyte isolation. Livers with severe steatosis, prolonged cold ischaemia time, older donors or other factors that make the tissue unsuitable for OLT are the main sources of human hepatocytes. The quality and viability of cells obtained from these livers are often poor and currently not sufficient for human hepatocyte transplantation.

Cell isolation can also be performed in remnants of the liver after orthotopic transplantation of reduced or split liver graft. Significant higher cell viability is obtained from these tissues when compared to those rejected for OLT *(10)*. Liver segment IV receives blood supply by the left hepatic artery and the left portal vein. When a liver is split between an adult and a paediatric patient, segment IV is allocated to the right lobe. At our centre, it is usually removed during the split procedures to avoid infarction and a potential risk of sepsis. In a study performed at our centre, three segments IV with or without the caudate lobe were used to isolate hepatocytes. From each segment about 0.5 billion hepatocytes were isolated, with a high viability of 90% *(11)*. Using these hepatocytes isolated from segment IV for clinical hepatocyte transplantation means that three patients can benefit from one split liver, effectively increasing the donor pool.

To increase the supply of tissues for OLT, non-heart-beating donors are being considered as an additional source of livers *(12)*. These organs are retrieved after the heart has stopped beating and respiration has ceased. As a result, liver tissues from this source have also become available for isolation of hepatocytes. A total of 20 livers or segments were perfused using the same methods as for the conventional donor livers, and the mean viability obtained was 52%. There was a significant negative correlation between hepatocyte viability and both warm and cold ischaemia periods. Only 35% of the livers processed achieved the viability required for clinical transplantation, which probably reflects that most of these livers had been rejected for whole-organ transplantation. The poor viability could be improved by reducing both cold and warm ischaemia times prior to processing *(13)*.

Other alternative sources of hepatocytes are being studied, such as immortalized cell lines *(14, 15)*, foetal hepatocytes *(16)* and stem cell-derived hepatocytes *(17–19)*, and will be discussed elsewhere in this book.

2.2. Isolation of Hepatocytes

There are well-established protocols for isolation of human hepatocytes *(10, 20)* based on the collagenase digestion of perfused liver tissue at 37°C. Once the liver tissue is digested and cells released, the hepatocytes are separated by low-speed centrifugation, and the pellets obtained are washed with ice-cold buffer solution to purify the cells. The cell viability and yield are then assessed, and will vary depending on the quality of the tissue used. Hepatocytes need to be used as soon as possible for cell transplantation, preferably within 24 h of isolation, as function deteriorates even when kept at 4°C. For longer-term storage of human hepatocytes, a number of cryopreservation protocols are available *(21)*. In most of them, hepatocytes are maintained at 4°C after isolation and cryopreserved as soon as possible. The best results are currently obtained by cryopreservation in a mixture of the organ preservation media University of Wisconsin solution and final concentration of 10% dimethyl sulphoxide (Me_2SO) using a controlled-rate cell freezer *(22)*. There are so many steps involved in hepatocyte isolation and cryopreservation that often insufficient viable hepatocytes are recovered on thawing. The cryopreserved hepatocytes can then be stored at –140°C until required for clinical use.

2.3. GMP Laboratory and Cell Banking

An aseptic environment is required to prepare cells on a large scale in conditions of good manufacturing practice (GMP), so that the isolated cells are safe to be administered to humans. The cell isolation unit is a purpose-built facility consisting of interconnected rooms. Air entering the laboratory passes through HEPA filters to remove any particles and an air-handling unit maintains a temperature-controlled environment inside the unit. There is a gradient of air pressures between the rooms, which maintains a positive air pressure differential, with the highest pressure in the aseptic room, where tissue processing is performed. Operators have to wear sterile clean-room suits. Standard operating procedures are followed for all aspects of work in the cell isolation unit. A comprehensive quality control system monitors all aspects of laboratory performance.

Cryopreserved hepatocytes for clinical use are stored in cell freezer bags in the vapour phase of liquid nitrogen inside an automated storage container. A cell bank permits the immediate use of hepatocytes in urgent cases of liver disease.

All donated organs/tissues should be screened for viral infection, including hepatitis and human immunodeficiency virus according to the National Solid Organ Transplant Service criteria. The final cell products must be screened for the presence of microorganisms. For clinical transplantation, hepatocytes must have a viability higher than 60%, a yield superior to 5×10^8 hepatocytes and the absence of microbiological contamination.

3. Pre-clinical Studies

Extensive laboratory studies in experimental animal models of human liver disease established the feasibility and efficacy of hepatocyte transplantation into various sites such as liver, spleen, pancreas, peritoneal cavity and sub-renal capsule. Identification of transplanted hepatocytes was documented by a number of different methods. Models have included the identification of normal hepatocytes transplanted into Nagase analbuminaemic or dipeptidyl peptidase IV-deficient rats by liver (immuno)histochemistry and serum albumin levels, in the case of Nagase analbuminaemic rats. Another approach used was the use of donor cells secreting or expressing unique reporter proteins, including the green fluorescent protein for direct identification of transplanted cells *(23, 24)*.

Hepatocyte transplantation improves the survival of animal models with ALF, induced either chemically *(25–27)* or surgically *(28)*. For human metabolic disorders, there are several animal models, including the Gunn rat (model for CN syndrome type I), the fumarylacetoacetate hydrolase–/– knockout mice (model of tyrosinaemia type I), the Long Evans Cinnamon rat (model of Wilson's disease), the mdr2 mouse (model of progressive familial intrahepatic cholestasis type 3), the spf-ash mouse (model of congenital ornithine transcarbamylase (OTC) deficiency), the

Watanabe heritable hyperlipidemic rabbit (model for LDL receptor deficiency) and the hyperuricemic Dalmatian dog. Hepatocyte transplantation showed improvement of the biochemical abnormalities in metabolic models, but complete correction of the genetic abnormalities required a significant amount of engrafted cells. Repeated hepatocyte transplantation can increase the number of engrafted liver cells *(29)*, although better results are seen in animal models where donor hepatocytes have a selective advantage over the native hepatocytes to repopulate the recipient liver *(30–32)*.

4. Clinical Hepatocyte Transplantation

4.1. Acute Liver Failure

Animal studies encouraged human clinical application of hepatocyte transplantation, initially in the treatment of patients with ALF. Eighteen patients who received hepatocyte transplantation for ALF, from six centres in the United States, were reviewed by Strom et al. *(33)*. Infusion of 10^7–10^9 hepatocytes, either fresh or after cryopreservation, was performed into the splenic artery or portal vein. Up to a maximum of 5% of normal liver mass was infused and it is questionable whether this is a sufficient quantity to replace the massive lost function in ALF. In these studies, a reduction in ammonia and bilirubin levels and improvements in hepatic encephalopathy levels were reported, but liver cell transplantation did not significantly affect the clinical outcome of these patients. **Table 1.1** summarizes the overall data on ALF patients treated with hepatocyte transplantation.

4.2. Liver-Based Metabolic Disorders

The cell requirement for transplantation may be lower in some inherited metabolic liver diseases where the aim is to replace a single deficient enzyme. The first patients to receive hepatocyte transplantation for treatment of an inherited liver-based metabolic disorder were five children with familial hypercholesterolaemia. After liver resection, autologous hepatocytes were isolated and transduced ex vivo with a retroviral vector carrying the human LDL receptor and then transplanted back into the patients. There was evidence of engraftment and over 20% reduction in LDL cholesterol documented in three of the five patients transplanted, but less than 5% of transgene expression in donor hepatocytes after 4 months *(34, 35)*. Since then, many other patients have been treated with hepatocyte allotransplantation to correct metabolic diseases. The overall experience of hepatocyte transplantation for treatment of liver-based metabolic disorders, mainly in children, is shown in **Table 1.2**.

Table 1.1.
Worldwide results of human hepatocyte transplantation for ALF

Aetiology	No. of patients	Effect/outcome	Reference
Paediatric			
Drug	3	Ammonia reduction, but death at day 2 and 7 post-transplant	Soriano et al. (*51*)
	1	Death 4 days after intraportal infusion of 1 billion cells	Strom et al. (*33*)
	1	Ammonia decrease with improvement of encephalopathy. OLT at day 1 post-hepatocyte transplant	Fisher at al. (unpublished)
Idiopathic	2	Reduction in ammonia and encephalopathy. Full recovery in one and successful bridging to OLT in the other patient	Soriano et al. (*51*)
	1	No clear benefits, whole-organ transplant required	Sterling et al. (*52*)
	1	Full recovery with intraperitoneal injection of foetal hepatocytes	Habibullah et al. (*53*)
Viral	1	Decrease in ammonia levels and encephalopathy. Intracranial hypertension at day 2	Fisher et al. (unpublished)
Adults			
Drug	1	Improvement in encephalopathy. Death from sepsis at day 35	Strom et al. (*33*)
	3	Intrasplenic injection of hepatocytes. Decrease in ammonia levels and encephalopathy in 2 patients, but death from sepsis at days 14 and 20, respectively. No benefit in the third patient, developing multisystem organ failure within 6 h	Bilir et al. (*3*)
	2	Successful bridging to OLT at days 2 and 10 after hepatocyte transplantation into the spleen	Fisher et al. (*54*)
	4	No improvement in the patient who received intrasplenic infusions of hepatocytes. Brain death at day 1. Reduction in ammonia levels and encephalopathy after intraportal hepatocyte transplantation, with full recovery without OLT in one patient. Sepsis at day 18 and mesenteric thrombosis at day 3 were the causes of death in the other two patients	Fisher et al. (unpublished)

Table 1.1.
(continued)

Aetiology	No. of patients	Effect/outcome	Reference
	5	Single intraperitoneal infusion of 6×10^7 foetal hepatocytes/kg. Two of the five patients treated showed reduction in ammonia levels and improvement in encephalopathy, with full recovery	Habibullah et al. (*53*)
Idiopathic	1	Intrasplenic infusion of hepatocytes with decrease in ammonia levels and encephalopathy, allowing OLT at day 5. Death from multisystem organ failure after 13 days	Sterling et al. (*52*)
	1	Intraportal transplantation of hepatocytes for treatment of Reye's syndrome. Reduction in blood ammonia, but no improvement in encephalopathy. Death at day 1 post-transplant	Fisher et al. (unpublished)
Mushroom poisoning	1	Full recovery after intraportal infusion of 4.9×10^9 hepatocytes. Immunosuppression stopped after 12 weeks	Schneider et al. (*4*)
Post-surgical	1	No clinical improvement after intrasplenic transplantation of hepatocytes for ALF due to a trisegmentectomy. Death at day 2	Strom et al. (*33*)
Viral	2	Herpes II and Hepatitis B Virus (HBV) induced ALF treated by intraportal and intrasplenic liver cell transplantation. No benefit in the first patient, death after 18 h. Decrease in ammonia with improvement in encephalopathy, but multiorgan system failure at day 52	Bilir et al. (*3*)
	3	Two patients with HBV ALF treated by hepatocyte transplantation. One received intrasplenic infusion and showed decrease in blood ammonia and improvement in encephalopathy. Successful bridging to OLT at day 3. No benefit in the other after intraportal infusion, OLT at day 1. Improvement in encephalopathy and reduction in ammonia levels was observed in the third patient with herpes II, treated with intrasplenic infusion of hepatocytes. Death from sepsis at day 5	Strom et al. (*33*)
	1	Full recovery in HBV + cocaine ALF after intrasplenic hepatocyte transplantation	Fisher et al. (*55*)
	1	Decrease in ammonia levels and improvement in encephalopathy after intraportal infusion of hepatocytes for treatment of HBV + lymphoma ALF. Death from multiorgan system failure at day 7	Fisher et al. (unpublished)
	1	No benefit seen with intraperitoneal infusion of hepatocytes for treatment of ALF due to HBV infection. Death after 13 h	Habibullah et al. (*53*)

Modified from Fisher et al. (2006) *Transplantation* 82, 441–449.

Table 1.2
Hepatocyte transplantation: clinical studies in liver-based metabolic diseases

Liver disease	No. of patients	Effect/outcome	Reference
Familial hypercholesterolaemia	5*	20% reduction in LDL cholesterol in 3 patients	Grossman et al. (*35*)
α1 AT deficiency	1	Intraportal infusion. OLT after 4 days. Cirrhosis on explanted liver	Strom et al. (*33*)
Crigler–Najjar syndrome type I	1	50% reduction in serum bilirubin	Fox et al. (*5*)
	2	40% reduction in serum bilirubin in one and no clear benefit in the other patient. Immunosuppression stopped after 5 months	Dhawan et al. (unpublished)
	1	Partial correction of clinical jaundice. OLT after 5 months due to a very poor quality of life	Ambrosino et al. (*6*)
	1	30% decrease in serum bilirubin and phototherapy requirement	Allen et al. (personal communication)
Factor VII deficiency	3	80% reduction in recombinant factor VII requirement	Dhawan et al. (*39*)
Glycogen storage disease type Ia	1	Normal diet with no hypoglycaemia	Muraca et al. (*7*)
	1	Normal glucose 6 phosphatase activity up to 7 months	Lee et al. (personal communication)
	1	Partial response	Sokal et al. (personal communication)
Infantile Refsum's disease	1	Partial correction of metabolic abnormality	Sokal et al. (*38*)
Progressive familial intrahepatic cholestasis	2	No clear benefit – fibrosis already present. OLT at 5 and 14 months, respectively	Dhawan et al. (unpublished)
Urea cycle defect	1	Some clinical improvement. Died after 42 days	Strom et al. (*36*)
	1	Lowered blood ammonia and increased protein tolerance	Horslen et al. (*8*)
	1	No hyperammonaemia and increase in serum urea under normal protein diet. Auxiliary liver transplant at 7 months of age	Mitry et al. (*11*)
	2	Decrease in ammonia levels and improvement in psychomotor development	Stephenne et al. (*9*, *37*)
	1	Ammonia and citrulline levels decreased up to 6 months post-transplantation	Lee et al. (personal communication)

*Ex vivo gene therapy of autologous hepatocytes.

One of the key early reports was from Fox et al. in 1998, who reported the case of a 10-year-old girl with CN syndrome type I treated with hepatocyte transplantation. There was a reduction in her bilirubin levels and hours of phototherapy, and an increase in measured bilirubin UDP-glucuronosyl transferase activity after liver cell transplantation. Excretion of bilirubin conjugates in bile persisted for 3.5 years after hepatocyte transplantation. However, clinical improvements were not enough to ameliorate her quality of life, and the patient decided to undergo orthotopic auxiliary liver transplantation 4 years after liver cell transplantation *(5)*. Subsequently, four other patients with CN type I were treated with hepatocyte transplantation, two of them at King's College Hospital. The two patients received a total of 4.3 and 1.5×10^9 both fresh and cryopreserved hepatocytes. In the first patient who received nine infusions over 2 weeks and a further infusion 3 months later, there was an encouraging sustained reduction in serum bilirubin. The second child received three infusions of hepatocytes over a period of 3 weeks. No clear benefit in bilirubin levels was observed, and immunosuppression was stopped 5 months after hepatocyte transplantation. The patient is now listed for whole-organ transplantation. Two other patients with severe unconjugated hyperbilirubinaemia and clinical diagnosis of CN type I were treated with an intraportal infusion of 7.5 and 1.5×10^9 hepatocytes each, with a reduction of bilirubin levels by 30–50%. Due to poor tolerability to nocturnal phototherapy, the first child underwent OLT *(6)* (Allen et al., personal communication).

Five patients with urea cycle disorders have received hepatocyte transplantation, three of them for OTC deficiency, one for argininosuccinate lyase deficiency and one for citrullinaemia. The first, a 5-year-old boy with OTC deficiency, showed some clinical improvement, but died with hyperammonaemia 42 days after liver cell transplantation *(36)*. The second infant with a severe OTC mutation showed biochemical and clinical improvement for a short period after injection of hepatocytes, but activity was lost, probably because of acute rejection *(8)*. Our first patient to receive hepatocyte transplantation was a 1-day-old boy with an antenatal diagnosis of severe OTC deficiency. Infusion of 1.6×10^9 hepatocytes was performed via an umbilical vein catheter. After transplantation, he had no episodes of hyperammonia and showed an increase in urea synthesis while on a normal protein diet. The child underwent auxiliary liver transplantation at 7 months of age due to uncertainties about the long-term efficacy of hepatocyte transplantation *(11)*. Liver cell transplantation was used as a bridge to OLT in a 14-month-old boy with OTC deficiency poorly equilibrated by conventional therapy. He was maintained on a restricted protein diet, sodium benzoate therapy and arginine/citrulline supplementation and received 3.5×10^9 cryopreserved cells into the portal vein (10 infusions over 16 weeks). Control of the ammonia levels and urea synthesis, as well as improved psychomotor development, was observed until OLT, 6 months after the first infusion of cells *(37)*. Recently, a 42-month-old girl with argininosuccinate lyase deficiency and secondary psychomotor retardation because of recurrent episodes of hyperammonaemia was treated with hepatocyte transplantation. Repeated intraportal injections of fresh and cryopreserved hepatocytes to reach 9% of her total hepatic mass were performed over 5 months. A metabolic and psychomotor improvement was observed, and there was evidence of hepatocyte engraftment up to 12 months after cell transplantation *(38)*. The last patient with urea cycle disorder to receive hepatocyte transplantation was a 25-month-old child with citrullinaemia. With intraportal hepatocyte transplantation of 10% of the calculated liver mass, a decrease in both ammonia and citrulline levels was achieved up to 6 months post-transplant (Lee et al., personal communication).

In two adults with glycogen storage disease type Ia, hepatocyte transplantation resulted in improved glucose control on a normal diet, and one of the patients showed normal glucose 6 phosphatase activity for 7 months (*7*) (Lee et al., personal communication). The only child to receive intraportal infusion of human hepatocytes as a treatment for this metabolic disease showed only partial response (Sokal et al., personal communication).

The first use of hepatocyte transplantation for treatment of inherited coagulation factor VII deficiency was at King's College London, in two brothers who presented a severe form of this condition. Both children received hepatocytes (a total of 1.1 and 2.2×10^9) through a Hickman line inserted in the inferior mesenteric vein. Infusion of isolated human hepatocytes improved the coagulation defect and markedly decreased the requirement for exogenous recombinant factor VIIa (rFVIIa) to around 20% of that before cell transplantation. Six months post-hepatocyte transplantation in both cases higher rFVIIa doses were required, suggesting the loss of transplanted hepatocyte function, possibly associated with sepsis. Due to increasing problems with venous access and uncertainty about the long-term efficacy of hepatocyte transplantation, OLT was performed successfully in both cases *(39)*. Subsequently, a third patient with factor VII deficiency received a total of 2.8×10^9 hepatocytes (fresh and cryopreserved) and showed similar outcome (Dhawan et al., unpublished).

Two other children treated in 2003 were suffering from progressive familial intrahepatic cholestasis (PFIC2), a genetic disease where the liver is lacking the bile salt export pump *(40)*. As a result of this defect, bile flow is severely impaired and patients rapidly develop liver cirrhosis and need liver transplantation. Both children with PFIC2 received a single percutaneous transhepatic injection of one-third of a billion fresh hepatocytes into the portal system. The rationale was that the injected hepatocytes would have a selective growth advantage over the defective host hepatocytes to repopulate the liver, as had been shown in a mouse model of progressive familial intrahepatic cholestasis type 3 *(30)*, where up to 70% of host hepatocytes were replaced by donor cells. However, both patients had a whole-liver transplant 5 and 14 months later, respectively, as their livers had continued to deteriorate. Existing fibrosis in the hepatic sinusoids is likely to have impaired engraftment of transplanted hepatocytes into the liver structure. Earlier treatment, if feasible, may be the best approach in this situation.

Among the other patients reported, a child with α1-antitrypsin deficiency was found to have cirrhosis at the time of cell infusion and underwent subsequent liver transplantation *(33)*. Finally, a child with infantile Refsum's disease had a partial correction in the metabolic abnormality after liver cell transplantation and persistent evidence of peroxisomal function up to 18 months later *(38)*.

4.3. Route of Administration

The liver and the spleen are the most consistent sites for hepatocyte engraftment and function. Intraportal injection is the preferred delivery method for clinical hepatocyte transplantation. The portal venous system can be accessed using different techniques: percutaneous transhepatic puncture of the portal vein, transjugular approach to the right portal vein, catheterization of the mesenteric vein or umbilical vein catheterization in newborn babies. Hepatic ultrasound and portal venous system Doppler examination should be performed before the procedure to exclude any malformation or venous thrombosis. The percutaneous transhepatic portal vein access technique was first described in 1967 by Aronsen and Nylander *(41)*. Since then the technique has been widely used for diagnostic portography, embolization procedures and, most recently, for cell transplantation. It can be performed under general anaesthesia or simple sedation combined to local anaesthetic agents. The potential complications associated with the percutaneous transhepatic approach are mainly hepatic haematoma, portal vein thrombosis, haemorrhage, puncture of the biliary system and vasovagal reactions *(42, 43)*. Combined ultrasound or computed tomography and fluoroscopy guidance have been performed in an attempt to reduce the number of punctures to gain access to the portal vein, thus decreasing the procedure-related risks *(42, 44)*. The transjugular approach to the right portal vein is another method to be considered for hepatocyte transplantation, but is more complex and cannot be performed under ultrasound guidance *(42)*. In any of these methods, the portal venous pressure must be carefully monitored throughout the procedure. Repeated cell infusions are normally required when a large amount of hepatocytes has to be injected.

To avoid multiple anaesthetic procedures and portal vein punctures, surgical placement of a long-term intravenous access in the mesenteric vein should be considered. The use of an implantable mesenteric Port-a-Cath® device was recently described as a practical means to infuse hepatocytes *(45)*.

The spleen is considered an adequate site for hepatocyte transplantation, particularly in cirrhotic patients. When injected into the splenic bulb, cells translocate to the liver through the splenic vein. Another attractive site for cell transplantation is the peritoneal cavity due to its large capacity and simple access. In spite of the fact that isolated hepatocytes do not normally engraft or survive following intraperitoneal injection, transplantation of encapsulated or matrix-attached hepatocytes has prolonged cell survival in animal models *(46)*.

4.4. Immuno suppression

To date there is no consensus regarding the immunosuppressive treatment, but most centres have used the protocol of liver transplantation. Combination of tacrolimus and steroids with or without sirolimus or mycophenolate mofetil has been used. Some centres use monoclonal antibodies like basiliximab or daclizumab. The Edmonton protocol for islet cell transplantation appears to be the most promising and our centre is beginning to follow this regimen.

5. The Future

Considerable progress has been made in bringing hepatocyte transplantation to the bedside. However, the success of hepatocyte transplantation from animal models experiments could not be fully reproduced in humans. Although results in clinical studies have been encouraging, no complete correction of any metabolic disease in patients by hepatocyte transplantation alone has been reported. There are still a number of areas for improvement and development.

The limited supply of livers currently available to isolate hepatocytes is a major problem for hepatocyte transplantation. As discussed before, donor liver tissues unsuitable for OLT are currently the principal source of human hepatocytes. Livers with moderate-to-severe steatosis are those most commonly rejected for clinical transplantation and represent an important potential source of hepatocytes. The improvement of the outcome of isolation and purification of these hepatocytes is an important goal, so that these cells could be used for transplantation. It is not likely that the supply of hepatocytes will increase, so a wider use of hepatocyte transplantation will not be possible until alternative sources of cells are found. Foetal hepatocytes, liver stem/

progenitor cells isolated from adult livers, embryos, umbilical cord blood and bone marrow, and hepatocytes conditionally immortalized by gene transfer are ongoing areas of investigation. There is a focus of research worldwide on liver stem cell biology and there is no doubt that there are many hurdles to cross before clinical application will be possible. Xenotransplants could be a potentially unlimited source of fresh hepatocytes; however, there are many concerns regarding rejection and transmission of infectious diseases that need to be resolved.

Another limiting factor of the technique is the conservation and storage of isolated cells. There is a need to improve the storage of hepatocytes, both for longer periods in the cold so they can be used fresh after a number of days and also better cryopreservation protocols for longer term storage. Viability and function on thawing of cryopreserved hepatocytes can be improved by the use of protocols incorporating cryo/cytoprotectant agents *(47)*.

The demonstration of engraftment and repopulation of the recipient liver by donor hepatocytes is still a major difficulty. In some liver-based metabolic disorders, the restoration of a metabolic defect after liver cell transplantation can be assessed from serum concentration of a metabolite, but this may not provide reliable information on the number of surviving and functioning engrafted cells. Moreover, the distribution of the engrafted cells cannot be determined by this approach. Other techniques require a liver biopsy to determine donor engraftment, such as short tandem repeats analysis *(48)*, quantitation of gene expression of liver-specific transcripts and fluorescence in situ hybridization *(9)* or real-time PCR of Y chromosome *(49)*, in cases of sex-mismatched hepatocyte transplantation. The disadvantages of hepatic biopsies are procedure-related morbidity and selective sampling of the graft at a single endpoint. For these reasons, reliable non-invasive methods are required to monitor cell survival and engraftment after transplantation. There is growing interest in using magnetic resonance imaging to track cells after in vitro labeling with contrast agents *(50)*.

It is also clear that many injected cells do not engraft into the recipient liver and are either cleared by the reticuloendothelial system or lose viability during this early phase. The outcome of hepatocyte transplantation would benefit from methods to enhance engraftment and repopulation by the induction of a selective growth advantage over host hepatocytes, although the options for this in humans would be limited. Rejection of the allogeneic hepatocytes and/or eventual senescence of the cells transplanted are probably contributing factors for the loss of long-term function of these cells in clinical transplants. More studies are needed to minimize or overcome the need of immunosuppression in liver cell transplantation. If this could be achieved, hepatocyte transplantation would exhibit an exceptional advantage over OLT.

In summary, considerable experience has been gained so far in the handling of hepatocytes and techniques for hepatocyte transplantation allowing clinical hepatocyte transplantation. This will give a good basis for the future application of new technologies, particularly those based on stem cells, which, it is hoped, will increase the utilization of cell transplantation.

Reference

1. Pereira, S. P., McCarthy, M., Ellis, A. J., et al. (1997) Auxiliary partial orthotopic liver transplantation for acute liver failure. *J Hepatol* 26, 1010–1017.
2. Strom, S. C., Fisher, R. A., Thompson, M. T., et al. (1997) Hepatocyte transplantation as a bridge to orthotopic liver transplantation in terminal liver failure. *Transplantation* 63, 559–569.
3. Bilir, B. M., Guinette, D., Karrer, F., et al. (2000) Hepatocyte transplantation in acute liver failure. *Liver Transpl* 6, 32–40.
4. Schneider, A., Attaran, M., Meier, P. N., et al. (2006) Hepatocyte transplantation in an acute liver failure due to mushroom poisoning. *Transplantation* 82, 1115–1116.
5. Fox, I. J., Chowdhury, J. R., Kaufman, S. S., et al. (1998) Treatment of the Crigler–Najjar syndrome type I with hepatocyte transplantation. *N Engl J Med* 338, 1422–1426.
6. Ambrosino, G., Varotto, S., Strom, S. C., et al. (2005) Isolated hepatocyte transplantation for Crigler–Najjar syndrome type 1. *Cell Transpl* 14, 151–157.
7. Muraca, M., Gerunda, G., Neri, D., et al. (2002) Hepatocyte transplantation as a treatment for glycogen storage disease type 1a. *Lancet* 359, 317–318.
8. Horslen, S. P., McCowan, T. C., Goertzen, T. C., et al. (2003) Isolated hepatocyte transplantation in an infant with a severe urea cycle disorder. *Pediatrics* 111, 1262–1267.
9. Stephenne, X., Najimi, M., Sibille, C., et al. (2006) Sustained engraftment and tissue enzyme activity after liver cell transplantation for argininosuccinate lyase deficiency. *Gastroenterology* 130, 1317–1323.
10. Mitry, R. R., Hughes, R. D., Aw, M. M., et al. (2003) Human hepatocyte isolation and relationship of cell viability to early graft function. *Cell Transpl* 12, 69–74.
11. Mitry, R. R., Dhawan, A., Hughes, R. D., et al. (2004) One liver, three recipients: segment IV from split-liver procedures as a source of hepatocytes for cell transplantation. *Transplantation* 77, 1614–1616.
12. Muiesan, P. (2003) Can controlled non-heart-beating donors provide a solution to the organ shortage? *Transplantation* 75, 1627–1628.
13. Hughes, R. D., Mitry, R. R., Dhawan, A., et al. (2006) Isolation of hepatocytes from livers from non-heart-beating donors for cell transplantation. *Liver Transpl* 12, 713–717.
14. Kobayashi, N., Fujiwara, T., Westerman, K. A., et al. (2000) Prevention of acute liver failure in rats with reversibly immortalized human hepatocytes. *Science* 287, 1258–1262.
15. Cai, J., Ito, M., Nagata, H., et al. (2002) Treatment of liver failure in rats with end-stage cirrhosis by transplantation of immortalized hepatocytes. *Hepatology* 36, 386–394.
16. Dan, Y. Y., Riehle, K. J., Lazaro, C., et al. (2006) Isolation of multipotent progenitor cells from human fetal liver capable of differentiating into liver and mesenchymal lineages. *Proc Natl Acad Sci USA* 103, 9912–9917.
17. Avital, I., Feraresso, C., Aoki, T., et al. (2002) Bone marrow-derived liver stem cell and mature hepatocyte engraftment in livers undergoing rejection. *Surgery* 132, 384–390.
18. Miki, T., Lehmann, T., Cai, H., et al. (2005) Stem cell characteristics of amniotic epithelial cells. *Stem Cells* 23, 1549–1559.
19. Ruhnke, M., Ungefroren, H., Nussler, A., et al. (2005) Differentiation of in vitro-modified human peripheral blood monocytes into hepatocyte-like and pancreatic islet-like cells. *Gastroenterology* 128, 1774–1786.

20. Strom, S. C., Dorko, K., Thompson, M. T., et al. (1998) Large scale isolation and culture of human hepatocytes, in (Franco, D., et al. ed.), *Îlots de Langerhans et hépatocytes: vers une utilisation therapeutique*, pp. 195–205. Les Editions INSERM, Paris.
21. Terry, C., Dhawan, A., Mitry, R. R., et al. (2006) Cryopreservation of isolated human hepatocytes for transplantation: State of the art. *Cryobiology* 53, 149–159.
22. Diener, B., Utesch, D., Beer, N., et al. (1993) A method for the cryopreservation of liver parenchymal cells for studies of xenobiotics. *Cryobiology* 30, 116–127.
23. Horslen, S. P., Fox, I. J. (2004) Hepatocyte transplantation. *Transplantation* 77, 1481–1486.
24. Fox, I. J., Roy-Chowdhury, J. (2004) Hepatocyte transplantation. *J Hepatol* 40, 878–886.
25. Krishna Vanaja, D., Sivakumar, B., Jesudasan, R. A., et al. (1998) In vivo identification, survival, and functional efficacy of transplanted hepatocytes in acute liver failure mice model by FISH using Y-chromosome probe. *Cell Transpl* 7, 267–273.
26. Sutherland, D. E., Numata, M., Matas, A. J., et al. (1977) Hepatocellular transplantation in acute liver failure. *Surgery* 82, 124–132.
27. Baumgartner, D., LaPlante-O'Neill, P. M., Sutherland, D. E., et al. (1983) Effects of intrasplenic injection of hepatocytes, hepatocyte fragments and hepatocyte culture supernatants on D-galactosamine-induced liver failure in rats. *Eur Surg Res* 15, 129–135.
28. Demetriou, A. A., Reisner, A., Sanchez, J., et al. (1988) Transplantation of microcarrier-attached hepatocytes into 90% partially hepatectomized rats. *Hepatology* 8, 1006–1009.
29. Rozga, J., Holzman, M., Moscioni, A. D., et al. (1995) Repeated intraportal hepatocyte transplantation in analbuminemic rats. *Cell Transpl* 4, 237–243.
30. De Vree, J. M., Ottenhoff, R., Bosma, P. J., et al. (2000) Correction of liver disease by hepatocyte transplantation in a mouse model of progressive familial intrahepatic cholestasis. *Gastroenterology* 119, 1720–1730.
31. Laconi, E., Oren, R., Mukhopadhyay, D. K., et al. (1998) Long-term, near-total liver replacement by transplantation of isolated hepatocytes in rats treated with retrorsine. *Am J Pathol* 153, 319–329.
32. Guha, C., Parashar, B., Deb, N. J., et al. (2002) Normal hepatocytes correct serum bilirubin after repopulation of Gunn rat liver subjected to irradiation/partial resection. *Hepatology* 36, 354–362.
33. Strom, S. C., Chowdhury, J. R., Fox, I. J. (1999) Hepatocyte transplantation for the treatment of human disease. *Semin Liver Dis* 19, 39–48.
34. Grossman, M., Raper, S. E., Kozarsky, K., et al. (1994) Successful ex vivo gene therapy directed to liver in a patient with familial hypercholesterolaemia. *Nat Genet* 6, 335–341.
35. Grossman, M., Rader, D. J., Muller, D. W., et al. (1995) A pilot study of ex vivo gene therapy for homozygous familial hypercholesterolaemia. *Nat Med* 1, 1148–1154.
36. Strom, S. C., Fisher, R. A., Rubinstein, W. S., et al. (1997) Transplantation of human hepatocytes. *Transpl Proc* 29, 2103–2106.
37. Stephenne, X., Najimi, M., Smets, F., et al. (2005) Cryopreserved liver cell transplantation controls ornithine transcarbamylase deficient patient while awaiting liver transplantation. *Am J Transpl* 5, 2058–2061.
38. Sokal, E. M., Smets, F., Bourgois, A., et al. (2003) Hepatocyte transplantation in a 4-year-old girl with peroxisomal biogenesis disease: technique, safety, and metabolic follow-up. *Transplantation* 76, 735–738.
39. Dhawan, A., Mitry, R. R., Hughes, R. D., et al. (2004) Hepatocyte transplantation for inherited factor VII deficiency. *Transplantation* 78, 1812–1814.
40. Thompson, R., Strautnieks, S. (2001) BSEP: function and role in progressive familial intrahepatic cholestasis. *Semin Liver Dis* 21, 545–550.
41. Aronsen, K. F., Nylander, G. (1967) Use of direct protography in diagnosis of liver diseases. *Radiology* 88, 40–47.
42. Goss, J. A., Soltes, G., Goodpastor, S. E., et al. (2003) Pancreatic islet transplantation: the radiographic approach. *Transplantation* 76, 199–203.
43. Maleux, G., Gillard, P., Keymeulen, B., et al. (2005) Feasibility, safety, and efficacy of percutaneous transhepatic injection of beta-cell grafts. *J Vasc Interv Radiol* 16, 1693–1697.
44. Owen, R. J., Ryan, E. A., O'Kelly, K., et al. (2003) Percutaneous transhepatic pancreatic islet cell transplantation in type 1 diabetes mellitus: radiologic aspects. *Radiology* 229, 165–170.

45. Darwish, A. A., Sokal, E., Stephenne, X., et al. (2004) Permanent access to the portal system for cellular transplantation using an implantable port device. *Liver Transpl* 10, 1213–1215.
46. Fox, I. J., Chowdhury, J. R. (2004) Hepatocyte transplantation. *Am J Transpl* 4 Suppl 6, 7–13.
47. Terry, C., Dhawan, A., Mitry, R. R., et al. (2005) Preincubation of rat and human hepatocytes with cytoprotectants prior to cryopreservation can improve viability and function upon thawing. *Liver Transpl* 11, 1533–1540.
48. Mas, V. R., Maluf, D. G., Thompson, M., et al. (2004) Engraftment measurement in human liver tissue after liver cell transplantation by short tandem repeats analysis. *Cell Transpl* 13, 231–236.
49. Wang, L. J., Chen, Y. M., George, D., et al. (2002) Engraftment assessment in human and mouse liver tissue after sex-mismatched liver cell transplantation by real-time quantitative PCR for Y chromosome sequences. *Liver Transpl* 8, 822–828.
50. Rogers, W. J., Meyer, C. H., Kramer, C. M. (2006) Technology insight: in vivo cell tracking by use of MRI. *Nat Clin Pract Cardiovasc Med* 3, 554–562.
51. Soriano, H. E., Wood, R. P., Kang, D. C. (1997) Hepatocellular transplantation in children with fulminant liver failure. *Hepatology* 30, 239A.
52. Sterling, R. K., Fisher, R. A. (2001) Liver transplantation: Living donor, hepatocyte, and xenotransplantation, in (Gish, R., ed.), *Current Future Treatment Therapies for Liver Disease. Clinics in Liver Disease*, WB Saunders, Philadelphia.
53. Habibullah, C. M., Syed, I. H., Qamar, A., et al. (1994) Human fetal hepatocyte transplantation in patients with fulminant hepatic failure. *Transplantation* 58, 951–952.
54. Fisher, R. A., Strom, S. C. (2000) Human hepatocyte transplantation: Biology and therapy, in (Berry, M. N., Edwards, A. M., ed.), *Hepatocyte Review*, Kluwer Academic Publishers, Dordrecht, The Netherlands.
55. Fisher, R. A., Bu, D., Thompson, M., et al. (2000) Defining hepatocellular chimerism in a liver failure patient bridged with hepatocyte infusion. *Transplantation* 69, 303–307.

Chapter 2

Isolation of Human Hepatocytes

Ragai R. Mitry

Abstract

Protocols for isolation of human hepatocytes have been developed. The isolated cells can be used not only in research but also for transplantation in patients with liver disease, especially acute liver failure and liver-based metabolic/synthetic conditions. The aim of hepatocyte transplantation is to correct the missing liver function(s) and allow either the recovery of the liver or buy the patient time until a suitable donor liver is available for transplantation.

Key words: Hepatocyte transplantation, donor liver, collagenase.

1. Introduction

Hepatocyte transplantation is emerging as a treatment for liver-based metabolic disease and as a means of liver support in acute liver failure patients *(1)*. The technique is dependent on the availability of good quality hepatocytes isolated from unused/rejected liver for transplantation on the grounds of being severely steatotic, or having a long cold ischaemia time, and also the remnants of liver after transplantation of a reduced size or split liver graft. The level of viability and cellular activity of isolated hepatocytes are dependent on the quality of the original tissue. The technique used for isolation of hepatocytes from liver tissue is a standard collagenase perfusion technique based on the original work by Berry and Friend *(2)*, which was later modified by Seglen *(3)*.

Anil Dhawan, Robin D. Hughes (eds.), *Hepatocyte Transplantation, vol. 481*

DOI 10.1007/978-1-59745-201-4_2 Springerprotocols.com

2. Materials

2.1. Human Liver Tissue

The following protocol can be used for isolation of hepatocytes from donor liver tissue unused/rejected for transplantation, and is based on previously published protocols *(4, 5)*. Appropriate ethical approvals and signed consent forms must be obtained prior to processing of any tissues, and the appropriate rules and regulations for human tissue processing, cell handling and storage must be followed (*see* **Note 1**).

2.2. Chemicals and Solutions

The following is a list of the chemicals and solutions used in the hepatocyte isolation procedure and cell culture:

1. Hank's Balanced Salt Solution (HBSS) without calcium or magnesium (Cat. No. 10-547F; Cambrex Bio Science Wokingham Ltd., Berkshire, UK)
2. Eagle's Minimum Essential Medium containing 25 mM HEPES (4-(2-hydroxyethyl)-1-piperazineethanesulphonic acid) (EMEM/HEPES), without phenol red and calcium (Cat. No. 12-136Q; Cambrex, UK)
3. 1 M HEPES solution (Cat. No. BE17-737E; Cambrex, UK)
4. Collagenase P (Cat. No. 11213873001; Roche Diagnostics Ltd., East Sussex, UK)
5. Ethyleneglycol-bis(beta-aminoethyl ether)-N,N,N',N'-tetraacetic acid (EGTA) (Cat. No. E4378; Sigma-Aldrich Company Ltd., Dorset, UK)
6. 1.0 N NaOH solution (Cat. No. 319511; Sigma-Aldrich Ltd.)
7. Bovine serum albumin (BSA), (Cat. No. A2153; Sigma-Aldrich Ltd.) (*see* **Note 2**)
8. DNaseI (Cat. No. DN25; Sigma-Aldrich Ltd.) (*see* **Note 3**)
9. William's E medium (WEM) (Cat. No. E7023; Sigma-Aldrich Ltd.)
10. Foetal calf serum (FBS), heat-inactivated (Cat. No. 10108-165; Invitrogen Ltd., Paisley, UK)
11. Insulin (Cat. No. I1882; Sigma-Aldrich Ltd.)
12. Dexamethasone (Cat. No. D8893; Sigma-Aldrich Ltd.)
13. Ethanol (Cat. No. E7023; Sigma-Aldrich Ltd.)
14. Glacial acetic acid (Cat. No. A9967; Sigma-Aldrich Ltd.)
15. QuantiChrom™ Urea Assay Kit (Cat. No. DIUR-500; BioAssay Systems, Hayward CA, USA)
16. 1× Phosphate-buffered saline (PBS) tablets (Cat. No. P4417; Sigma-Aldrich Ltd.)
17. Sterile deionised water
18. Distilled water

2.3. Preparation of Solutions

2.3.1. Perfusion Solutions

The four buffer solutions required during the isolation and preparation of human hepatocytes are listed below. Sufficient volumes of these solutions must be prepared under sterile conditions (i.e. inside a cell culture laminar flow cabinet).

1. 250 mM EGTA: dissolve 1.902 g EGTA in 1.0 N NaOH solution (final volume should be 20 ml) and sterilise by filtration using a 0.2 μm filter inside a laminar flow cabinet. The EGTA solution should be stored as small aliquots in a fridge.
2. Perfusion solution 1 (P1): for every 500 ml HBSS add 1 ml of 250 mM EGTA stock solution and ~2.3 ml 1 M HEPES and mix well (final pH should be 7.3–7.4).
3. Perfusion solution 2 (P2): 500 ml HBSS.
4. Perfusion solution 3 (P3): 1 l EMEM/HEPES containing 0.5 g collagenase P. Collagenase should be weighed in a sterile Falcon® tube and dissolved in 50 ml of the EMEM/HEPES. The collagenase solution is sterilised by passing it through a 0.2 μm filter into a fresh 50 ml Falcon® tube, then add to the 950 ml EMEM/HEPES and mix well.
5. Wash solution (W): 1 l EMEM/HEPES containing 50 g BSA (final concentration 5%). BSA should be weighed, dissolved and sterilised prior to use similar to collagenase preparation (see step 3 above). Maintain the sterile solution on ice until required.

Perfusion solutions (P1, P2 and P3) must be maintained at 37°C after preparation, while the wash solution (W) should be maintained on ice.

Example: for 100 g liver tissue prepare 500 ml of P1, 500 ml of P2, 1000 ml of P3 and 1000–1200 ml of W.

2.3.2. Preparation of Supplements and Culture Medium

1. Dexamethasone (40 μg/ml): dissolve 1 mg dexamethasone in 1 ml absolute alcohol (ethanol) by gentle swirling, then add 24 ml culture medium. Store solution in small aliquots at –20°C. Avoid repeated freeze/thaw.
2. Insulin (10 mg/ml): dissolve 100 mg insulin in 10 ml acidified sterile water (pH ≤ 2.0; prepared by adding approx. 0.1 ml glacial acetic acid to 9.9 ml water). Store solution in small aliquots at 2–8°C (stable for 1 year).

The culture medium to be used should be prepared by adding the following supplements to 500 ml of WEM:

- 50 ml FBS
- 5 ml of 1 M HEPES
- 5 ml L-glutamine

- 5 ml penicillin/streptomycin
- 0.5 ml of dexamethasone stock solution
- 28.6 µl of insulin stock solution

Mix well by gentle swirling of the medium bottle. The bottle could be stored at 2–8°C for up to 1 month.

2.4. Other Items

1. Water bath.
2. Multi-channel perfusion pump (e.g. Masterflex® L/S Pump purchased from Cole-Parmer Instrument Company Ltd., London, UK).
3. Perfusion tubes: Masterflex® silicon rubber tubings size 16 (Cat. No. 96400-16; Cole-Parmer Instrument Company Ltd.). Short pieces (10 cm approximately) of this tubing are used for cannulating blood vessels (*see* **Note 4**).
4. Bottle top works to fit perfusion solution bottles (Cat. No. 734-5043; VWR International, Leicestershire, UK). It is a bottle cap with three tubes passing through it.
5. Sterile swabs (10×10 cm) type Topper 8 (Cat. No. TS8105; Johnson & Johnson Medical, Skipton, UK).
6. Refrigerated benchtop centrifuge.
7. Short connectors (Avon Medicals Cat. No. R93; SIMS Portex Ltd., Kent, UK).
8. Sutures (3-0 or 4-0), e.g. Ethicon-coated Vicryl® (Cat. No. W9130; Johnson & Johnson Medical).
9. BD BioCoat™ collagen I 24-well multiwell plates (Cat. No. 356408; BD Biosciences, San Jose CA, USA).
10. Flat-bottom 96-well plates with lids (Cat. No. 734-2097; VWR International).

3. Methods

3.1. Liver Tissue Digestion

1. Major blood vessels on the cut surface of the liver tissue are cannulated and the cannulae secured by suturing. Other small blood vessels not used for perfusion should be closed by suturing to minimise fluid leakage during perfusion.
2. A short connector is fitted to the free end of each cannula.
3. Long perfusion tubes are passed through the perfusion pump heads, and using a short connector, connect one of the free ends of each perfusion tube to the bottle top works fitted to the P1 solution bottle, which is maintained in the water bath at 37°C.

4. The perfusion lines are then primed with P1 solution. The other free end of each perfusion tube is then connected to the short connector fitted to the cannula.
5. The perfusion pump is then set to 60–80 ml/min flow rate and then switched on to start the perfusion process.
6. Following the perfusion with the three perfusion solutions (P1, P2 and P3), the digested tissue is then transferred into a sterile metal bowl. The cannulae and sutures are removed, and ice-cold W solution is poured onto the digested tissue until the tissue is completely covered.
7. Mince digested tissue using a sterile pair of scissors or scalpel blades to release hepatocytes, followed by filtration through two single layers of sterile swabs.

3.2. "Purification" of Hepatocytes

1. Aliquot the cell suspension obtained into 50 ml Falcon® tubes, and pellet hepatocytes by centrifugation at $50\times g$, 4°C for 4 min.
2. Discard supernatant, then resuspend each pellet in 50 ml ice-cold W solution, and re-centrifuge tubes. Repeat the wash/centrifugation steps two to three more times.
3. Estimate the cell count and viability using the standard Trypan blue exclusion technique *(6)*.
4. Fresh hepatocytes are ready to use, or cryopreserved and stored in the vapour phase of liquid nitrogen storage tank or in a –140°C freezer for future use (*see* **Chapter 3**).

3.3. Synthetic/Metabolic Activity Assay

Several liver- or hepatocyte-specific functional assays could be used to assess or evaluate the synthetic/metabolic activity of the isolated hepatocytes such as the production of urea resulting from the detoxification of ammonia.

3.3.1. Urea Production

Urea could be measured in the culture medium of hepatocyte cultures. The isolated hepatocytes are plated in wells of collagen-coated 24-well plates, and after 24 h incubation, samples of the culture medium are collected and analysed (*see* **Note 5**).

3.3.1.1. Plating Hepatocytes

1. Place 1 ml PBS in each well of the collagen-coated culture plate, and incubate the plate in a cell culture incubator for 10–15 min.
2. Remove PBS and place 3×10^5 hepatocytes in each well in 500 µl WEM with supplements.
3. Incubate the plate for 24 h in the cell culture incubator.
4. Collect culture medium samples from all wells, and measure the urea levels.

3.3.1.2. Measurement of Urea in Culture Medium

This assay is carried out according to the supplier's protocol.

1. Dilute the urea standard (50 mg/dl) provided in the kit to a final concentration of 10 mg/dl. This could be done by mixing

Table 2.1
Urea standard curve dilutions

Final urea concentration (mg/dl)	Volume of diluted urea standard (μl)	Volume of distilled water (μl)
0	0	50
2	10	40
4	20	30
6	30	20
8	40	10
10	50	0

80 μl urea standard with 320 μl distilled water in a 1.5 ml microfuge tube.

2. Use the diluted urea standard (10 mg/dl) to prepare a urea standard curve with a range of 0–10 mg/dl (**Table 2.1**).
3. Place the urea standards in duplicates of 50 μl in the wells of a flat-bottom 96-well plate.
4. Place duplicates of 25 μl culture medium samples in the wells and add 25 μl distilled water to each well. A duplicate of fresh sample of culture medium/distilled water (1:1) must be included, and its mean urea value must be subtracted from the mean urea values of the test samples (*see* **Note 6**).
5. Prepare enough "working reagent" by mixing equal volumes of Reagent A and Reagent B (provided in the kit) shortly prior to assay.
6. Add 200 μl of "working reagent" per well and tap the plate lightly to mix.
7. Cover the plate and incubate for 30 min at room temperature.
8. Read optical density at 470–550 nm (peak absorbance at 520 nm) using a plate reader.
9. Using the urea standard curve, estimate the levels of urea in your samples. Urea values of the culture medium samples must be multiplied by the dilution factor of 2.

4. Notes

1. Clinical-grade hepatocytes can be prepared under strict sterile conditions using the same isolation protocol. This requires the processing of the liver tissue and hepatocytes in an accredited

good manufacturing practice unit, which operates according to regulations set by a specialised governmental agency, e.g. in the United Kingdom, the Human Tissue Authority (*see* **Notes 2 and 3**).

2. BSA is an animal product and must not be used if the cells isolated from unused donor liver tissue and are going to be used for clinical transplantation. Human serum albumin should be used instead.
3. Over-digestion of the perfused tissue leads to an increased number of dead cells, which may release their contents. One of the released components is DNA, which is "sticky" and acts as "glue", making cells stick together with the formation of cell clumps. To avoid this problem, DNaseI could be added (50 mg/l) to solution P3 at the time of preparation. DNaseI must not be used if cells are going to be used for clinical transplantation.
4. For narrow blood vessels intravenous cannula (16–18 G) could be used.
5. Culture medium samples could be stored at –20°C until required for analysis.
6. FCS added to the culture medium contains urea and may affect the results; therefore a duplicate of samples of diluted fresh culture medium must be analysed alongside the test samples.

References

1. Fisher, R. A., Strom, S. C. (2006) Human hepatocyte transplantation: worldwide results. *Transplantation* 82, 441–449.
2. Berry, M. S., Friend, D. S. (1965) High yield preparation of isolated rat liver parenchymal cells. *J Cell Biol* 43, 506–520.
3. Seglen, P. O. (1976) Preparation of rat liver cells. *Meth Cell Biol* 13, 29–83.
4. Strom, S. C., Dorko, K., Thompson, M. T., et al. (1998) Large scale isolation and culture of human hepatocytes, in (Franco, D., Boudjema, K., Varet, B., eds.), *Îlots de Langerhans et hépatocytes*, pp. 195–205. Les Editions INSERM, Paris.
5. Mitry, R. R., Hughes, R. D., Aw, M. M., et al. (2003) Human hepatocyte isolation and relationship of cell viability to early graft function. *Cell Transpl* 12, 69–74.
6. Freshney, R. I. (2000) *Culture of Animal Cells*. Wiley-Liss, New York, NY, pp. 309–328.

Chapter 3

An Optimised Method for Cryopreservation of Human Hepatocytes

Claire Terry and Robin D. Hughes

Abstract

Successful cryopreservation of hepatocytes is essential for their use in hepatocyte transplantation. Cryopreservation allows hepatocytes to be available for emergency treatment of acute liver failure and also for planned treatment of liver-based metabolic disorders. In addition, cryopreservation of human hepatocytes can facilitate their use in metabolism and toxicity studies. Cryopreservation can adversely affect the viability and function, especially reduce the attachment efficiency, of hepatocytes on thawing.

The cryopreservation process can be divided into steps so that improvements can be made on the 'standard' protocols that are followed in some laboratories. These steps are as follows: pre-incubation of cells; freezing solution, cryoprotectants and cytoprotectants; freezing process; storage; thawing; post-thawing culture. This chapter presents an optimised protocol for cryopreservation of human hepatocytes as developed at King's College Hospital.

Key words: Human hepatocytes, cryopreservation, freezing, hepatocyte function, UW solution, glucose, fructose.

1. Introduction

Human hepatocyte preparations are limited by a lack of human tissue. Sources (from rejected or unused donor tissue or from liver resection tissue) are limited, erratic and unpredictable. However, when tissue is available, often large numbers of cells can be isolated. The problem is that usually not all the cells can be used immediately and hepatocytes do not proliferate in vitro *(1)*. Therefore, a reliable method for preserving hepatocytes is essential. Currently, the only method for long-term preservation of cells is cryopreservation.

Hepatocyte cryopreservation was first fully investigated and published in the 1980s *(2, 3)*. Since then cryopreservation protocols

Anil Dhawan, Robin D. Hughes (eds.), *Hepatocyte Transplantation, vol. 481*

DOI 10.1007/978-1-59745-201-4_3 Springerprotocols.com

have been published for hepatocytes from a variety of animal species, including rat *(4, 5)*, pig *(6, 7)*, mouse *(8, 9)*, monkey *(10, 11)* and dog *(12, 13)*. Optimised human hepatocyte cryopreservation protocols are fewer, presumably due to the limitation of human tissue to prepare hepatocytes for experiments, but there are still a large number of published human protocols *(3, 14–22)*. Even with the best of these protocols, there is still a significant loss of function and this is related to the quality of the fresh cells and the type and nature of the liver tissue from which they were isolated *(23)*. The state of the art of cryopreservation for hepatocyte transplantation has recently been reviewed *(24)*.

2. Materials

2.1. Pre-incubation

1. William's E Medium (WEM, Sigma-Aldrich Company Ltd., Gillingham, Dorset, UK) is prepared with the following additions: penicillin (50 U/ml, Life Technologies Ltd., Paisley, Scotland, UK) and streptomycin (50 μg/ml, Life Technologies Ltd.), L-glutamine (2 mM, Life Technologies Ltd.) and 4-(2-hydroxyethyl)-1-piperazineethanesulphonic acid (HEPES, 100 μM, Sigma-Aldrich Company Ltd.).
2. Heat-inactivated foetal calf serum (FCS, Life Technologies Ltd.).
3. Falcon tubes – 50 ml sterile conical bottom (BD Biosciences, Cowley, Oxfordshire, UK).
4. Glucose, fructose, α-lipoic acid (Sigma-Aldrich Company Ltd.).

2.2. Freezing Solution

1. University of Wisconsin (UW) solution (Bristol-Myers Squibb Pharma Ltd., Hounslow, UK).
2. Dimethyl sulphoxide (DMSO, Sigma-Aldrich Company Ltd.).

2.3. Cryopreservation Process

1. Kryo 10 Controlled Rate Freezer (CRF), Series III (Planer Products Ltd., Middlesex, UK).
2. Cryotubes (5 ml, Nunc Nalgene, Hereford, UK).

2.4. Storage

1. –140°C freezer (Lab Impex Research Ltd., East Sussex, UK).

2.5. Thawing

1. Waterbath (Model JB2, Grant Instruments (Cambridge) Ltd., Royston, Hertfordshire, UK).

2.6. Culture and In Vitro Cell Assays

1. Trypan blue solution (0.4%, Sigma-Aldrich Company Ltd.).
2. Collagen-coated (Biocoat™) flat-bottom 96-well culture plates (BD Biosciences).
3. Culture media consists of phenol red-free WEM with the additions in **Section 2.1**, Point 1, and 5% (v/v) FCS.

4. An incubator for culture is used (95% O_2/5% CO_2, Function-Line Incubator, Heraeus Instruments, Hanau, Germany).
5. For subsequent in vitro assays of hepatocyte function, serum-free WEM is used (i.e. the above WEM without the FCS addition).

3. Methods

3.1. Pre-incubation

1. Human hepatocytes (1.5×10^7 cells/tube) isolated as described in this volume in **Chapter 2** by Mitry are pelleted by centrifugation at 50× *g* for 5 min at 4°C and the supernatant is removed.
2. The pellet is resuspended in 5 ml pre-incubation media consisting of WEM containing 10% FCS (*see* **Note 1**) and a pre-incubation compound (200 mM glucose, 200 mM fructose or 2.5 mM α-lipoic acid, *see* **Note 2**) to give a final cell density of 3×10^6 viable hepatocytes per millilitre in Falcon tubes (total of 1.5×10^7 cells in 5 ml pre-incubation media, *see* **Note 3**).
3. The pre-incubation tubes are then placed in a 4°C refrigerator for 2 h.

3.2. Freezing Solution

1. After 2 h of incubation, treatment tubes are mixed by inversion and centrifuged at 50× *g* for 5 min at 4°C.
2. The supernatant is removed and cryovials are kept on ice while the freezing solution is added.
3. Freezing media consists of UW solution (*see* **Note 4**). A concentration of 300 mM of glucose or 300 mM of fructose can also be added to the freezing solution. All freezing media should be freshly prepared on the day of use, and the pH checked and changed to pH 7.4 if necessary.
4. The freezing media is added, ice-cold, to the cryovials containing the hepatocyte pellets to make up the final volume (cells + freezing media) of 4.5 ml.
5. A volume of 0.5 ml DMSO (*see* **Note 5**) is then added, drop-wise, to all cryovials to give a final DMSO concentration of 10% (v/v).
6. The suspension can be kept on ice for a maximum of 5 min before the cryopreservation process begins.

3.3. Cryopreservation Process

1. The CRF should be set up, ensuring there is sufficient liquid nitrogen in the tank for the run, so that it is ready to begin freezing as soon as possible, or within 5 min, after the DMSO has been added to the hepatocyte solution.

Table 3.1
Optimised Controlled Rate Freezer Protocol

Step	Start temperature	Rate	Time	End temperature
1	8°C	-1°C/min	8 min	0°C
2	0°C	HOLD	8 min	0°C
3	0°C	-2°C/min	4 min	-8°C
4	-8°C	-35°C/min	33 s	-28°C
5	-28°C	-2.5°C/min	2 min	-33°C
6	-33°C	+2.5°C/min	2 min	-28°C
7	-28°C	-1°C/min	32 min	-60°C
8	-60°C	-10°C/min	4 min	-100°C
9	-100°C	-20°C/min	2 min	-140°C

2. When the CRF has reached the start temperature (8°C), samples are inserted into the tube rack and the freezing protocol initiated (*see* **Note 6**).
3. **Table 3.1** shows the standard freezing protocol used, consisting of nine steps (*see* **Notes 7 and 8**).
4. The freezing protocol takes approximately 50 min.

3.4. Storage

1. The frozen cryovials should be immediately transferred to a −140°C freezer (*see* **Notes 9 and 10**).

3.5. Thawing

1. After storage at −140°C (*see* **Note 11**), the frozen cell suspensions can be rapidly thawed in a 37°C water bath with gentle agitation (*see* **Note 12**).
2. When all ice has disappeared (~1–2 min), the cell suspension can be transferred to a fresh ice-cold tube.
3. Dilution of the cryoprotectant should be carried out immediately with thawing media consisting of ice-cold WEM containing 20% FCS and an additional cytoprotectant if required (300 mM glucose, 300 mM fructose or 5 mM α-lipoic acid).
4. For every 1 ml of cell suspension thawed, the following volume of thawing medium is added drop-wise and with 5 min on ice between each addition: 0.5, 1, 2, 3 and 6 ml *(19)*.

5. The hepatocytes are then pelleted by centrifugation at $50\times g$ at 4°C for 5 min and the pellet is resuspended in a known volume of WEM.
6. For a description of modifications of the protocol for clinical hepatocytes (*see* **Note 13**).

3.6. Culture and In Vitro Cell Assays

1. Cell counts and crude viability assessments can be determined using the trypan blue exclusion method.
2. Hepatocytes are cultured (30,000 viable cells/well) in 96-well flat-bottomed collagen-coated plates.
3. Culture media consists of WEM containing 10% FCS, penicillin (50 U/ml) and streptomycin (50 μg/ml), and L-glutamine (2 mM) at 37°C in 95% O_2/5% CO_2.
4. After 24 h of culture, attachment efficiency can be determined by measuring the protein content *(25)* of attached cells and that of the initial number of cells (30,000 total cells/well).

4. Notes

1. FCS is commonly used as an addition to cell culture media to provide a 'cocktail' of factors required for cell proliferation and maintenance *(26)*. The complex list of components in FCS includes growth factors (e.g. epidermal growth factor, platelet-derived growth factor), trace elements (e.g. iron, zinc), lipids (e.g. cholesterol, linoleic acid), polyamines (e.g. putrescine, ornithine), attachment factors (e.g. fibronectin, laminin), mechanical protection and buffering capacity (e.g. albumin), metal transporters (e.g. transferin, ceruloplasmin) and protease inhibitors (e.g. α_1-antitrypsin, α_2-macroglobulin). 10% FCS is often used as an addition to WEM for culture of hepatocytes. It can also be used in cryopreservation media but cannot be used for cryopreserving hepatocytes for clinical transplantation.
2. Pre incubation of hepatocytes with glucose, fructose or α-lipoic acid at 4°C prior to cryopreservation has been found to improve thawed hepatocyte viability and function *(27)*. There was no evidence that using the three compounds in combination had an additive effect.
3. It is possible to successfully freeze cells at densities of up to 1×10^7/ml, if larger cell numbers are required, for clinical use. For this purpose, 50 and 250 ml Cryocyte freezing bags (Baxter, Oxford, UK) may be more suitable than cryotubes.
4. UW solution is an intracellular fluid type electrolyte composition with high potassium and low sodium content. The solution aims to improve hypothermic storage by five mechanisms:

(1) minimising hypothermic-induced cell swelling; (2) preventing intracellular acidosis; (3) preventing the expansion of interstitial space; (4) preventing injury from oxygen free radicals; and (5) providing substrates for regenerating high-energy phosphate compounds (e.g. ATP). These aims are achieved by including lactobionate (to prevent cell swelling and acidosis) and trisaccharide raffinose (to increase osmotic pressure). Hydroxyethyl starch and raffinose elevate the intracellular osmotic pressure to stabilise the cell membrane. Mannitol is a hydroxyl radical scavenger. Glutathione, adenosine and allopurinol facilitate the production of ATP and prevent active oxygen-induced cellular damage.

5. DMSO is able to enter the cells (a permeable cryoprotectant) and reduces cell injury by moderating the increase in solute concentration during freezing. The polar sulphoxide moiety of DMSO also interacts electrostatically with phospholipid membranes *(28)*. DMSO has been shown to decrease the temperature at which the lamellar phases of phosphatidylethanolamines are induced to transform into hexagonal-II structures (non-lamellar structure) that preserve membrane integrity during freeze-thaw *(29)*. During freezing, DMSO can keep the non-bilayer lipids in an association with intrinsic membrane proteins and prevent phase separation of the non-bilayer lipids during the cooling phase *(30)*.
6. To monitor the temperature changes in the cell suspension, an extra cryovial containing the standard cell suspension can be used with the CRF temperature probe inserted to record the temperature during freezing.
7. The aim of the CRF protocol is to attain a linear decrease of temperature in the cell suspension during freezing (**Fig. 3.1**).
8. The CRF protocol introduces a shock cooling step at the point when crystallisation is estimated to occur, to prevent the latent heat of fusion, which is suddenly released at the point of

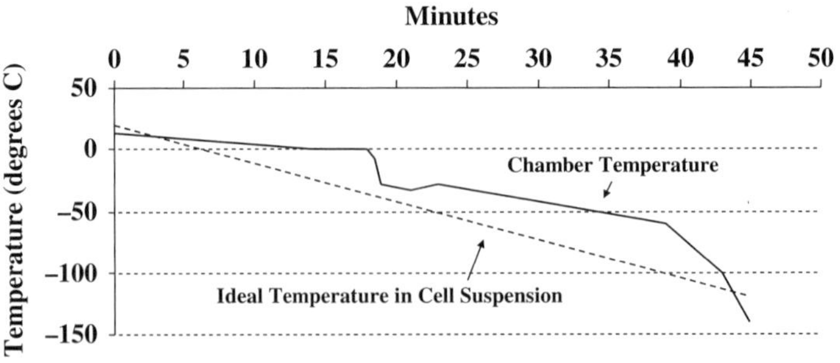

Fig. 3.1. Standard Freezing Protocol Employed with the Controlled Rate Freezer. The actual temperature decrease in the freezing chamber of the CRF (*solid line*) and the desired temperature decrease in the cell suspension (*dashed line*) according to the standard freezing protocol are shown. The freezing protocol employs different rates of freezing to try and attain this linear decrease in the cell suspension temperature.

crystallisation resulting in the cell sample being warmed slightly. This phenomenon has also been investigated in rat hepatocytes by Houle et al. *(31)*, who showed that the release of latent heat occurred at –29°C with an increase of 2°C. By introducing this shock cooling step at –8°C (rapid cooling from –8°C to –28°C in 6 s), controlled nucleation of ice and immediate crystallisation of the cell suspension were achieved and the damaging latent heat release eliminated. An additional step (increase in temperature to –28°C in 2 min) to prevent too rapid cooling of the cell suspension complemented the protocol. The protocol also takes advantage of the strategy to avoid cryopreservation damage by using rapid cooling interrupted with steps of isothermal holding periods to achieve enough cellular dehydration to prevent intracellular ice formation while minimising the total freezing time. The period of holding the hepatocytes at 0°C for 8 min allows time for transmembrane water transport. This approach minimises the cell exposure time to solution effects while avoiding the critical states associated with ice formation *(32, 33)*.

9. The storage temperature of cryopreserved hepatocytes is important. Storage at –80°C, for example, gives loss of cryopreserved human hepatocyte viability and cellular GSH content progressively over 1–4 days of storage after cryopreservation *(34)*.
10. Acceptable storage temperatures are at –140°C in a freezer, –150°C in the vapour phase of liquid nitrogen storage tanks (e.g. *14, 21, 35, 36)* or at –196°C in the liquid phase of liquid nitrogen (e.g. *15, 19, 22)*.
11. The possible length of storage time is debatable. Our study found no effect on hepatocyte viability or function after up to 3 years of storage. Generally, no effect of storage time is seen when hepatocytes are stored at <–140°C *(15, 18, 35)*. Dou et al. *(15)* have shown the successful storage of cryopreserved human hepatocytes for up to 1 year of storage at –196°C and Chesne et al. *(37)* have found the viability and attachment efficiency of cryopreserved human hepatocytes were unchanged after four years of storage in these conditions.
12. The optimum thawing protocol for hepatocytes is generally agreed to be rapid thawing at 37°C (to prevent recrystallisation) with slow dilution of the cryoprotectant (to reduce osmotic imbalances) at 4°C (to reduce possible toxicity of the cryoprotectant). If the thawing rate is not rapid enough, intracellular ice crystals can reform and coalesce into larger, more damaging crystals *(38, 39)*. If the cryoprotectant is not diluted out of the cell suspension slowly enough, osmotic shock may occur due to outflow of the cryoprotectant from the cells *(40)*. If the temperature of dilution is not at 4°C, toxicity may occur from further exposure to the cryoprotectant.

13. The protocol can be slightly adapted, using only clinically approved solutions, no animal products and following MHRA guidelines, to allow the hepatocytes to be suitable for clinical use. The following adaptations of the protocol are required:

- Pre-incubation may not be possible due to the larger number of cells to be cryopreserved. However, if this step is required the pre-incubation media should consist of UW solution containing either 300 mM glucose or 300 mM fructose only.
- Freezing bags (10 ml) can be used instead of cryovials with a freezing density of 1×10^7 viable cells/ml.
- Freezing media and thawing media should consist of UW solution with 10% DMSO and either glucose (300 mM) or fructose (300 mM).

Acknowledgements

We thank Merck Sharp and Dohme Ltd. and the Children's Liver Disease Foundation for their financial support.

References

1. Strom, S. C., Fisher, R. A., Rubinstein, W. S., et al. (1997) Transplantation of human hepatocytes. *Transpl Proc* 29, 2103–2106.
2. Fuller, B. J., Morris, G. J., Nutt, L. H., et al. (1980) Functional recovery of isolated rat hepatocytes upon thawing from –196°C. *Cryo Lett* 1, 139–146.
3. Loretz, L. J., Li, A. L., Flye, M. W., et al. (1989) Optimization of cryopreservation procedures for rat and human hepatocytes. *Xenobiotica* 19, 489–498.
4. Chesne, C., Guillouzo, G. A. (1988) Cryopreservation of isolated rat hepatocytes: a critical evaluation of freezing and thawing conditions. *Cryobiology* 25, 323–330.
5. De Loecker, W., Koptelov, V. A., Grischenko, V. I., et al. (1998) Effects of cell concentration on viability and metabolic activity during cryopreservation. *Cryobiology* 37, 103–109.
6. Koebe, H. G., Dahnhardt, C., Muller-Hocker, J., et al. (1996) Cryopreservation of porcine hepatocyte cultures. *Cryobiology* 33, 127–141.
7. Naik, S., Santangini, H. A., Trenkler, D. M., et al. (1997) Functional recovery of porcine hepatocytes after hypothermic or cryogenic preservation for liver support systems. *Cell Transpl* 6, 447–454.
8. Canaple, L., Nurdin, N., Angelova, N., et al. (2001) Maintenance of primary murine hepatocyte functions in multicomponent polymer capsules – *in vitro* cryopreservation studies. *J Hepatol* 34, 11–18.
9. Nyberg, S. L., Sreekumar, R., Yagi, T., et al. (2001) Impact of cryopreservation on hepatocyte gene expression. *ILTS ELTA LICAGE Berlin* 63.
10. De Sousa, G., Nicolas, F., Placidi, M., et al. (1999) A multi-laboratory evaluation of cryopreserved monkey hepatocyte functions for use in pharmaco-toxicology. *Chemico-Biol Interact* 121, 77–97.
11. Hewitt, N. J., Fischer, T., Zuehlke, U., et al. (2000) Metabolic activity of fresh and cryopreserved cynomolgus monkey (Macaca fasciculris) hepatocytes. *Xenobiotica* 30, 665–681.
12. Kasai, S., Mito, M. (1993) Large-Scale Cryopreservation of isolated dog hepatocytes. *Cryobiology* 30, 1–11.
13. Swales, N., Utesch, D. (1998) Metabolic activity of fresh and cryopreserved dog hepatocyte suspensions. *Xenobiotica* 28, 937–948.

14. Rijintes, P. J. M., Moshage, H. J., Van Gemert, P. J. L., et al. (1986) Cryopreservation of adult human hepatocytes: the influence of deep freezing storage on the viability, cell seeding, survival, fine structures and albumin synthesis in primary cultures. *J Hepatol* 3, 7–18.
15. Dou, M., De Sousa, G., Lacarelle, B., et al. (1992) Thawed human hepatocytes in primary culture. *Cryobiology* 29, 454–469.
16. Diener, B., Traiser, M., Arand, M., et al. (1994) Xenobiotic metabolising enzyme activities in isolated and cryopreserved human liver parenchymal cells. *Toxicol In Vitro* 8, 1161–1166.
17. Adams, R. M., Wang, M., Crane, A. M., et al. (1995) Effective cryopreservation and long-term storage of primary human hepatocytes with recovery of viability, differentitation, and replicative potential. *Cell Transpl* 4, 570–586.
18. Li, A. P., Gorycki, P. D., Hengstler, J. G., et al. (1999) Present status of the application of cryopreserved hepatocytes in the evaluation of xenobiotics: consensus of an international expert panel. *Chemico-Biol Interact* 121, 117–123.
19. Steinberg, P., Fischer, T., Kiulies, S., et al. (1999) Drug metabolizing capacity of cryopreserved human, rat, and mouse liver parenchymal cells in suspension. *Drug Metab Disp* 27, 1415–1422.
20. Hengstler, J. G., Utesch, D., Steinberg, P., et al. (2000) Cryopreserved primary hepatocytes as a constantly available *in vitro* model for the evaluation of human and animals drug metabolism and enzyme induction. *Drug Metab Rev* 32, 81–118.
21. Ostrowska, A., Bode, D. C., Pruss, J., et al. (2000) Investigation of functional and morophological integrity of freshly isolated and cryopreserved human hepatocytes. *Cell Tissue Bank* 1, 55–68.
22. Alexandre, E., Viollon-Abadie, C., David, P., et al. (2002) Cryopreservation of adult human hepatocytes obtained from resected liver biopsies. *Cryobiology* 44, 103–113.
23. Terry, C., Mitry, R. R., Lehec, S. C., et al. (2005) The effects of cryopreservation on human hepatocytes obtained from different sources of liver tissue. *Cell Transpl* 14, 527–536.
24. Terry, C., Dhawan, A., Mitry, R. R., et al. (2006) Cryopreservation of isolated human hepatocytes for transplantation: state of the art. *Cryobiology* 53, 149–159.
25. Lowry, O. H., Roseburgh, N. J., Farr, A. L., et al. (1951) Protein measurement with the Folin phenol reagent. *J Biol Chem* 193, 265–275.
26. Cartwright, T., Shah, G. P. (1994) Culture media. in (Davis, J. M., ed.), *Basic Cell Culture: A Practical Approach*, pp. 57–91. Oxford University Press, New York:.
27. Terry, C., Dhawan, A., Mitry, R. R., et al. (2006) Pre-incubation of rat and human hepatocytes with cytoprotectants prior to cryopreservation can improve viability and function on thawing. *Liver Transpl* 12, 165–177.
28. Anchordoguy, T. J., Cecchini, C. A., Crowe, J. H., et al. (1991) Insights into the cryoprotective mechanism of dimethyl sulfoxide for phospholipid bilayers. *Cryobiology* 28, 467–473.
29. Yu, Z. W., Quinn, P. J. (1998) The modulation of membrane structure and stability by dimethyl sulphoxide (Review). *Mol Memb Biol* 15, 59–68.
30. Quinn, P. J. (1985) A lipid phase separation model of low-temperature damage to biological membranes. *Cryobiology* 22, 128–146.
31. Houle, R., Raoul, J., Levesque, J. F., et al. (2003) Retention of transporter activities in cryopreserved, isolated rat hepatocytes. *Drug Metab Disp* 31, 447–451.
32. Harris, C. L., Toner, M., Hubel, A., et al. (1991) Cryopreservation of isolated hepatocytes: intracellular ice formation under various chemical and physical conditions. *Cryobiology* 28, 436–444.
33. Fuller, B. J., DeLoecker, L. W. (1997) Hepatocyte cryopreservation. in Mito, M., Sawa, M., eds.), *Hepatocyte Transplantation*, pp. 22–33. Karger Landes Systems, Netherlands.
34. Coundouris, J. A., Grant, M. H., Engeset, J., et al. (1993) Cryopreservation of human adult hepatocytes for use in drug metabolism and toxicity studies. *Xenobiotica* 23, 1399–1409.
35. De Sousa, G., Langouet, S., Nicolas, F., et al. (1996) Increase of cytochrome P-450 1A and glutathione transferase transcripts in cultured hepatocytes from dogs, monkeys, and human after cryopreservation. *Cell Biol Toxicol* 12, 351–358.
36. Skett, P., Roberts, P., Khan, S. (1999) Maintenance of steroid metabolism and hormone responsiveness in cryopreserved dog, monkey, and human hepatocytes. *Chemico-Biol Interact* 121, 65–76.

37. Chesne, C., Guyomard, C., Fautrel, A., et al. (1993) Viability and function in primary culture of adult hepatocytes from various animal species and human beings after cryopreservation. *Hepatology* 18, 406–414.
38. Karlsson, J. O. M., Toner, M. (1996) Long-term storage of tissues by cryopreservation: critical issues. *Biomaterials* 17, 243–256.
39. Karlsson, J. O. M., Cravalho, E. G., Borel Rinkes, I. H. M., et al. (1993) Nucleation and growth of ice crystals inside cultured hepatocytes during freezing in the presence of dimethylsulphoxide. *Biophys J* 65, 2524–2536.
40. Pegg, D. E. (2002) The history and principles of cryopreservation. *Semin Reprod Med* 20, 5–13.

Chapter 4

Liver Cell Culture Techniques

José V. Castell and María José Gómez-Lechón

Abstract

Different sources of hepatic tissue, including whole or split livers from organ donors or from cadavers, waste liver from therapeutic hepatectomies or small-sized surgical biopsies, can be successfully used to prepare human hepatocytes cultures. The two-step collagenase perfusion remains the most effective way to isolate high yields of viable hepatocytes from human liver samples that express many typical hepatic functions, among them drug-metabolising (detoxification) enzymes, when placed in primary culture. Once isolated, human hepatocytes cultured in monolayer in chemically defined conditions (serum-free) survive for limited periods of time gradually losing their differentiated phenotype, in particular the drug-metabolising enzymes. Supplementation of chemically defined media with growth factors, hormones and other specific additives has been used with variable success to extend hepatocyte survival and functionality in culture. Other culture improvements include the use of extracellular components to coat plates or to entrap cells. Conditions for short-term monolayer cultures, allowing the maintenance of liver-specific functions for approximately 1 week, are now well established. Cultures on plastic dishes coated with extracellular matrix components (i.e. Matrigel^{TM}, collagen, fibronectin or mixture of collagen and fibronectin) do meet many of the requirements for short-term incubation experiments, without adding too much complexity to the system. Practical details on how to carry out these cultures and to assess their functionality (CYP activity and ureogenesis) are discussed in this chapter.

Key words: Human hepatocytes, cell culture, collagen, fibronectin, ECOD, ureogenesis.

1. Introduction

Human hepatocytes are recognised as a closest model to human liver *(1, 2)*. Hepatocytes in chemically defined culture conditions express most typical hepatic biochemical functions, among which is the ability to metabolise drugs *(3–5)*. Primary hepatocytes are differentiated cells able to reproduce in vitro the response of human liver to chemicals and are currently considered a valuable in vitro tool for investigating drug metabolism *(6)* and

Anil Dhawan, Robin D. Hughes (eds.), *Hepatocyte Transplantation, vol. 481*

DOI 10.1007/978-1-59745-201-4_4 Springerprotocols.com

bioactivation and for assessing the potential hepatotoxicity of new drugs in man *(4, 5, 7)*.

1.1. Major Sources of Tissue Suitable for Human Hepatocyte Culture

Human liver tissue has become more available to many labs due, in part, to the expansion of the liver transplantation programmes. Different types of hepatic tissue, including non-implanted liver grafts (i.e. steatosis or non-identification of an adequate recipient), split livers from organ donors *(8)*, waste liver from therapeutic hepatectomies *(9, 10)* and, more recently, liver from non-heart-beating donors *(11)* have been successfully used to prepare human hepatocyte cultures *(12)*. The two-step collagenase perfusion remains the most effective way to isolate high yields of viable hepatocytes from human liver samples that express many typical hepatic functions, among them drug-metabolising (detoxification) enzymes, when placed in primary culture *(13)*.

The suitability of liver samples from different origins as a source of viable and metabolically competent human hepatocytes is variable *(14)*. Factors related to the procurement of the liver sample (warm and cold ischaemia) as well the intrinsic characteristics of liver tissue sample (sex, age, liver pathology, xenobiotic treatment, etc.) clearly influence the success of cultures *(4, 5, 15)*.

A comparative analysis of hepatocyte cultures from different types of liver tissue samples carried out in our laboratory is presented in **Table 4.1** *(5, 16, 17)*. Liver samples are grouped in three different categories: (a) *surgical biopsy* samples resected in the course of surgical procedures, not directly related to malignant pathology in hepatocytes; (b) liver samples obtained in the course of *partial hepatectomies* of liver tissue with a malignant tumoural process (hepatoma or metastasis) and (c) non-implanted *liver grafts* (tissue discarded for transplantation or remaining after size reduction). The key features to explore the quality and suitability of the tissue source were as follows: (1) cell yield; (2) viability of isolated cells, and cell protein attached to culture plates after 24 h of culture and (3) drug biotransformation capability of cultured hepatocytes (**Table 4.2**). Differences in cell viability and survival of cultured cells as related to age or gender of donors were much less relevant than the source and procurement of the tissue. Only well-preserved and rapidly processed tissues grant hepatocytes forming stable and functional monolayers (**Fig. 4.1**).

Elective surgical biopsies are by far the best quality source for hepatocytes *(15)*. The procedure of how samples are usually obtained, rapidly cooled and processed ensures high cellular yields and high viability and metabolic function (**Table 4.1**).

Therapeutic hepatectomy is another source of liver tissue for hepatocyte isolation. In contrast to other procedures to obtain liver samples, the hepatectomy technique requires clamping of vessels that irrigate the area of resection, which causes warm ischaemia of the hepatic tissue. The lower yields of viable

Table 4.1
Isolation and culture of human hepatocytes obtained by perfusion of liver samples from different sources

Source of liver sample	Yield ($\times10^{-6}$) (viable cells/gram liver)	Viability (%)	Successful cultures (%)	Cell protein (mg/plate)	ECOD (pmol/mg/min)	6β-OHT (pmol/mg/min)
Surgical biopsy ($n = 107$)	14.2±11	92±8	93	0.99±0.27	17.2±7.6	88.7±60.8
Hepatectomy ($n=18$)	8.2±5.8	91±7	77	0.90±0.17	17.3±5.5	81.5±54.4
Liver grafts ($n = 37$)	5.9±5.6*	57±30*	62	0.89±0.45	12.9±10.2	54.8±56.0

ECOD, ethoxycoumarin *O*-deethylase *(48)*; 6β-OHT, testosterone 6β-hydroxylase *(48)*. *$p<0.05$

hepatocytes obtained from therapeutic liver resections can be attributable to the fact that, in contrast to the other procedures to obtain liver samples, the partial hepatectomy requires clamping of vessels that irrigate the area prior to resection, presumably resulting in cell stress, the triggering of apoptosis (unpublished results) and reduced metabolic capability of hepatocytes (**Table 4.1**).

Livers from organ donors are usually perfused in situ with a cold preservation solution and kept under these conditions for several hours until processed for hepatocyte isolation. Cold ischaemia is a risk factor for organ function *(18)* and presumably influences the efficiency of the isolation procedure and the metabolic competence of cultured cells. Preservation solutions do have an influence on hepatocyte isolation. In our hands, Wisconsin solution *(18)*, because of its higher density and high raffinose content *(19)*, was much less suitable for hepatocyte isolation than Celsior *(16, 17, 19)*.

Table 4.2
Individual P450 activities in human hepatocytes prepared from human liver

Culture	MROD CYP1A2	COH CYP2A6	D4OH CYP2C9	M4OH CYP2C19	C6OH CYP2E1	6b-OHT CYP3A4
Mean ($n = 30$)	1.04	36.2	84.9	29.2	175	53.8
SD	0.63	20.5	40.7	27.2	111	40.1

Data are expressed as picomole of product formed per minute and per milligram of total cell protein.

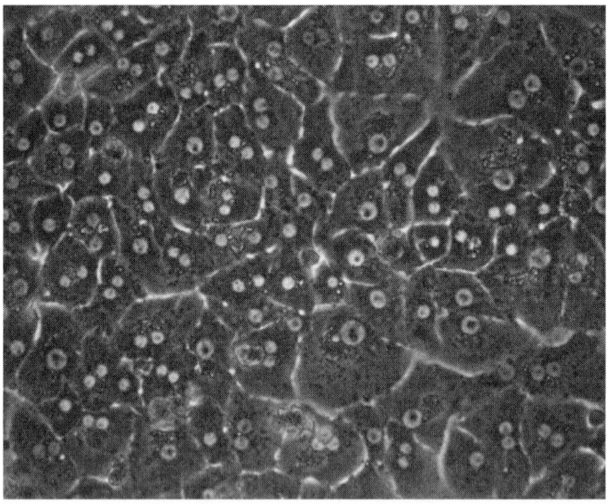

Fig. 4.1. Twenty-four hours primary cultured human hepatocytes in chemically defined media.

Steatosis is one of the major causes of donor organ refusal for transplantation, which then may become available for hepatocyte isolation. Steatosic livers of approximately 40–60% (pathologist confirmation) lead to a significant reduction in cell isolation yield, cell viability and function but still may be suitable for cell isolation. Steatosis >60% makes the liver tissue fully inappropriate for cell isolation *(20, 21)*.

1.2. Culture Media Composition

Human hepatocytes can adapt well to serum-free culture conditions. Significant advances have been made in prolonging cell survival and preserving liver-specific function in cultured hepatocytes by sophistication of culture media composition. Medium formulation influences the morphology, cell survival and functionality of hepatocytes in culture *(13, 17, 22, 23)*. Supplementation of chemically defined/serum-free media with growth factors and hormones *(22, 24, 25)*, or inhibitors of nitric oxide synthesis *(26)*, antioxidants *(17, 27)* and caspase inhibitors *(28)*, among others, have been used with success in an attempt to preserve hepatocyte functionality as well as long-lasting cultures *(29)*.

1.3. Culture Configuration

Once cells have been enzymatically isolated from the liver and placed in culture. The spatial configuration of cultures has a clear influence on cell survival and performance. Different culture techniques have been used to mimic in vitro the microenvironment of a hepatocyte in the liver by using plates coated with extracellular matrix components, synthetic ligands, co-cultures of hepatocytes with other cells, as well as three-dimensional (3D) cultures in biocompatible matrices, hepatocyte spheroids, etc.

Two-dimensional monolayer cultures. Improvements in two-dimensional (2D) monolayer cultures to favour functional maintenance of hepatocytes include the coating of culture plates with collagen *(27)*, fibronectin *(1)*, mixture of collagen and fibronectin *(1)* or Matrigel *(30)*. Cells adhere tightly as 2D monolayer cultures, but the tightly anchored hepatocytes do not remain fully phenotypically stable *(31, 32)*. Nevertheless, hepatocytes cultured on plastic dishes coated with extracellular matrix components do meet many of the metabolic features needed in most short-term incubation experiments, and thus represent a valuable and well-performing cellular model without adding too much complexity to the whole culture system. Practical details on how to carry out these cultures and to assess their functionality (CYP activity and ureogenesis) are given below.

3D cultures. The extracellular matrix is an important modulator of cell polarity and function, and influences the phenotype of both hepatocytes and non-parenchymal cells in the liver. The importance of reconstructing the extracellular matrix spatial geometry in hepatocyte cultures was first recognised by Dunn *(33)*. Sandwiching primary hepatocytes as monolayers within two layers of extracellular matrix is aimed at imitating its bilateral presence with respect to the sinusoidal surfaces of the hepatocytes (space of Disse). Extracellular matrix within the space of Disse next to the central vein is predominantly composed of collagen type I, and this protein, together with other extracellular matrix proteins (i.e. laminin, fibronectin), modulates hepatocyte growth, gene expression and stability of liver-specific functions *(14, 34–36)*. The morphological distinction between hepatocytes seeded onto collagen-coated plates without a collagen gel overlay (conventional monolayers) and those seeded onto collagen gel with a subsequent collagen gel overlay (sandwich) is visible just few hours after seeding. Conventional monolayer hepatocytes quickly adopt their polygonal shape and establish extensive cell–cell contacts, whereas in sandwich culture this takes markedly longer. In general, conventional monolayers appear more flattened than sandwich-cultured cells, a result of the lack of a 3D extracellular matrix environment. After overnight incubation, sandwich culture hepatocytes form aggregates with a typical cuboidal shape. Cells cultured as a collagen sandwich in a serum-free medium do not visibly spread out, and polygonal cell formats, clear plasma membrane boundaries and stable bile canaliculi-like networks *(35, 37)*. Microencapsulation in alginate, a relatively inert biocompatible matrix, also mimics the biological extracellular matrix, allowing the 3D configuration culture to cultivate successfully human hepatocytes *(14, 38)*.

Co-cultures. Heterotypic interactions between cells and non-parenchymal neighbours have been reported to modulate cell growth, migration and/or differentiation. In both the developing and adult liver, cell–cell interactions are imperative for coordinated organ function. In vitro, co-cultivation of hepatocytes and non-parenchymal cells has been used to stabilise the adult hepatocyte phenotype. Although the precise mechanisms by which non-parenchymal cells act on the hepatocyte phenotype remain non-elucidated, some new insights on the mode of cell signalling, cell–cell interaction and the ratio of cell populations are noted. Human hepatocytes co-cultured with an epithelial cell line derived from rat liver survived for more than 2 months and secreted high levels of albumin even in a serum-free medium. This long-term survival appeared to correlate with the production of an extracellular material, which is rich in collagen Type III *(39)*.

Spheroid cultures. Microaggregates of liver cells have been successfully established in an attempt to retain in vitro the type of cellular interactions that are likely to occur in the liver. Among the cell types incorporated into the culture aggregates are parenchymal and non-parenchymal liver cells (Kupffer, endothelial cells). Several reports indicate that culturing hepatocytes as multicellular aggregates maintain a prolonged expression of liver-specific genes and achieve polarity and cell-to-cell contact, resulting in upregulation of function *(40, 41)*. Spheroid culture systems favour the 3D cellular organisation and avoid the constraints of cell attachment support *(40)*. Spheroids of human hepatocytes have been reported to be viable up to at least 1 month in culture where they express a high cell functional hepatic activity *(30, 42)*.

Cultures on plastic dishes coated with extracellular matrix components (i.e. Matrigel™, collagen, fibronectin or mixture of collagen and fibronectin) do meet many of the requirements for short-term incubation experiments, without adding too much complexity to the system. Practical details on how to carry on these cultures, starting from collagenase isolated cells, as well to assess their functionality (CYP activity and ureogenesis) are described below

2. Materials

2.1. Reagents

1. *Enzymes:* Helix pomatia β-glucuronidase (EC. 3.2.1.31)/arylsulphatase (EC 3.1.6.1) preparations were obtained from Roche Diagnostics Corp.
2. *Chemicals:* Inorganic compounds were obtained from Sigma-Aldrich Chemicals. Sodium acetate, 7-ethoxycoumarin, 7-hydroxycoumarin, antipyrin and diacetylmonoxime were from Sigma-Aldrich Chemicals, as well.
3. *Ureogenesis:* Solution A: 19.6 mM antipyrin, 8.8 mM ferric ammonia sulphate, 4.51 M H_2SO_4 and 3.67 M H_3PO_4; solution B: 0.4% diacetylmonoxime in 7.5% w/v NaCl solution.

2.2. Culture Media

1. *Cell seeding culture medium:* Ham's F-12/Williams (1:1 v/v) medium (Gibco BRL, Paisley, Scotland) supplemented with 2% newborn calf serum (Gibco BRL), 0.1% bovine serum albumin (BSA) fraction V (Sigma, Madrid, Spain), 10 nM insulin (Novo Nordisk, A/S Bagsvaerd, Denmark), 25 μg/ml transferrin, 0.1 μM sodium selenite, 65.5 μM ethanolamine, 7.2 μM linoleic acid, 7 mM glucose, 6.14 mM ascorbic acid, 0.64 mM N^{omega}-nitro-L-arginine methyl ester (Sigma), and 50 mU/ml penicillin and 50 μg/ml streptomycin (Gibco BRL).
2. *Chemically defined culture media*: The same culture media composition as above, serum-free and supplemented with 10 nM dexamethasone (Sigma).
3. *Coating mixture for culture plates:* Prepare 100 ml of DMEM supplemented with 0.1% BSA fraction V. Dissolve 1 mg human fibronectin (Sigma) in 97 ml of DMEM supplemented with 0.1% BSA fraction V. Add to the former solution 3 ml of 0.1% collagen Type from calf skin solution in 0.1 M acetic acid.

3. Methods

3.1. Coating of Culture Plates

1. Culture plates are coated with 10 μl/cm^2 of the fibronectin/collagen coating mixture described above and allow to stand for 1 h.
2. Excess of coating mixture is removed and hepatocytes are seeded onto the plates at the appropriate density (*see* **Section 3.3**).

3.2. Cell Counting and Viability Assessment

1. After re-suspension of the cell pellet resulting from centrifugation of collagenase-digested liver tissue in the seeding culture medium, viability has to be determined to adjust cell seeding density.
2. 0.4% Trypan blue in saline is added to a diluted aliquot of the cell suspension and few microliters are immediately loaded inside a cell counter chamber.
3. Viable cells (colourless) and non-viable cells (deep blue) are counted in at least five different optical fields under the light microscope. Cell viability may vary between 90 and 60% depending on the origin and handling of the sample. Cell preparations with viability below 50% are generally not suitable for further cultivation and should be discharged.

3.3. Culture of Human Hepatocytes

Cell density influences the morphology of hepatocytes in culture *(43)*: when seeded at a very high density, cells do not spread out significantly; rather, they start to detach from the surface after 1–2 days.

1. The hepatocyte suspension is adjusted to a density of 5×10^3 cells/ml in seeding culture medium.

2. Hepatocytes are seeded on fibronectin/collagen-coated plates at a final density of 80×10^3 cells/cm^2 in an appropriate volume of culture medium.
3. One hour after cell seeding, the medium is aspirated to remove unattached cells and cell debris. Fresh culture medium is added to plates.
4. The attachment efficiency of hepatocyte suspension to fibronectin/collagen-coated plates is usually 80% of viable cells *(3, 44)*.
5. Twenty-four hours after cell plating, cells are shifted to serum-free/chemically defined culture medium (*see* **Section 2.1**) (**Fig. 4.1**).
6. Culture medium (*see* **Section 2.1**) is further renewed every following day.
7. Under these conditions, hepatocytes easily survive up to 7 days.

3.4. Quality-Control Assessment of Cultured Human Hepatocytes

3.4.1. Xenobiotic Metabolism Competence of Hepatocytes

The ability of hepatocytes to biotransform xenobiotics is one of the most relevant characteristics of differentiated hepatocytes, and needs to be examined in culture to ensure the appropriate metabolic performance of cells. The 7-ethoxycoumarin *O*-de-ethylation (ECOD) is catalysed by several CYP isoforms *(45)*, rendering fluorescent 7-hydroxycoumarin that can be easily monitored. The ECOD activity assay can be performed in intact cells and is a good indicator of global P450 activities. Because of it simplicity and consistency, the ECOD activity assay should be routinely measured in each hepatocyte culture preparation as quality control, in particular for those studies addressing drug metabolism and/or bioactivation-mediated cytotoxicity.

1. Seed hepatocytes in plastic culture dishes (typically, 3.5 cm diameter), as described in **Section 3.3**.
2. Twenty-four hours after cell seeding, wash plates twice with warm phosphate-buffered saline (PBS). Initiate the enzymatic assay by adding warm serum-free/chemically defined culture media containing 800 μM 7-ethoxycoumarin.
3. Incubate cells for about 45–60 min at 37°C; stop the reaction by aspirating the incubation medium from plates.
4. The enzymatically formed 7-hydroxycoumarin may be partially conjugated by cells to a less fluorescent derivative. To hydrolyse conjugates, add to a test tube 1 ml sample of cells' incubation media and 200 μl of a mixture of 200 Fishman units of β-glucuronidase and 1600 Roy units of arylsulphatase (Roche Diagnostics Corp), in an appropriate hydrolysis buffer (0.1 M sodium acetate pH 4.5); incubate for 2 h at 37°C.

5. Hydrolysis is stopped by adding 1 ml chloroform and vigorously shaking the mixture for 5–10 min.
6. After centrifugation (2000×*g* for 10 min), 0.6 ml of the organic phase is extracted with 1.2 ml of 1 M NaCl/0.01 N NaOH by vigorous vortexing.
7. Following centrifugation of the aqueous phase (2000×*g*, 10 min), the fluorescence is measured at 340 nm excitation and 460 nm emission in a fluorimeter.
8. A calibration curve is prepared by adding increasing amounts (0–5000 pmol/ml) of 7-hydroxycoumarin to chemically defined culture media. Standards are treated as described for regular samples (chloroform and NaOH extracted; *see* **Steps 4–7**).
9. A control (blank sample) is prepared adding 800 μM 7-ethoxycoumarin to a chemically defined culture medium. The blank is processed as described for assay samples (*see* **Steps 4–7**) and the fluorescence measured.
10. Fluorescence values of samples are corrected by subtracting the blank reading. The amount of 7-hydroxycoumarin formed by hepatocytes is calculated by interpolation of the corrected fluorescence in the standard curve (*see* **Step 8**).
11. Plates, after aspirating the incubation medium (*see* **Step 3**), are washed once with PBS, and the protein content is measured by the Lowry or Bradford methods. The protein content of plates is used to normalise ECOD activity values.
12. The activity is expressed as picomoles of 7-hydroxycoumarin formed per minute and per milligram of cell protein.
13. Typical ECOD activities in 24-h cultured human hepatocytes range 16.5±8.4 pmol of 7-hydroxycoumarin per minute per milligram of cell protein ($n = 88$).

3.4.2. Ureogenesis

Ureogenesis from ammonia occurs exclusively in the liver. Urea synthesis involves both cytosolic and mitochondrial reactions *(46)* and is a valuable global indicator of hepatic performance and of the degree of mitochondrial preservation. Under basal conditions, human hepatocytes in a monolayer culture synthesize urea at a rate of 2.5–3.5 nmol per mg cell protein per min *(1, 4)*. Urea synthesis in the human liver can be estimated in approximately 1.2 nmol per mg cell protein per min *(47)*. Twenty-four hour cultured hepatocytes can be maximally stimulated with ammonia to synthesise up to 130 nmol urea per mg cell protein per min *(1, 4)*.

The following procedure has been adapted for an easy measurement of ammonia-stimulated urea production by hepatocytes in culture and allows a rapid and convenient assessment of their metabolic performance:

1. Seed hepatocytes in petri culture plates (3.5 cm diameter) as described in **Section 3.3**. After 24 h of culture, wash plates twice with warm PBS. Initiate the assay by adding 1.5 ml of 3 mM NH_4Cl dissolved in serum-free/chemically defined media.
2. Incubate cells for 2 h at 37°C. Withdraw 200 μl aliquots of incubation medium every 30 min.
3. To each aliquot add 1.5 ml of a reaction milieu containing a 2:1 (v/v) mixture of solutions A and B.
4. Mix thoroughly and incubate samples for 15 min at 100°C in a water bath in the dark. Reaction develops a green colour in the samples. To stop the reaction, cool down the samples.
5. A standard curve is prepared by adding increasing amounts of urea (200–400 nmol/1.7 ml of reaction buffer) in 200 μL of chemically defined culture media. Standards are treated as described for samples (*see* **Steps 4–6**).
6. A control (blank sample) is taken out of chemically defined/serum-free culture media containing 3 mM ammonia, and treated as described (*see* **Steps 4–6**).
7. The absorbance of the samples, standards and blank are read at 464 nm in a spectrophotometer.
8. Absorbance values are corrected by subtracting the blank readings. The urea formed by hepatocytes is calculated by interpolation of the corrected absorbance in the standard curve (*see* **Step 5**).
9. Culture plates are washed once with PBS, and the protein content is measured by the Lowry or Bradford methods. The protein content of plates is used to normalise urea production by cells.
10. The urea production rate is usually expressed as nanomoles of urea formed per minute and per milligram of cell protein.

Acknowledgements

The authors are indebted to Generalitat Valenciana and Foundation Lubasa for their support in the creation of the Unit of Cell Transplantation. This research is part of CIBERHED, and was supported by EU grants "Predictomics" and "Carcinogenomics".

References

1. Gomez-Lechon, M. J. (1997) *In isolation, culture and use of human hepatocytes in drug research: In Vitro Methods in Pharmaceutical Research*, Academic Press, London.
2. Maurel, P. (1996) The use of adult human hepatocytes in primary culture and other *in vitro* systes to investigate drug metabolism in man. *Adv Drug Deliv Rev* 22, 105–132.

3. Gomez-Lechon, M. J. (2000) *Isolation and culture of human hepatocytes. In: The Hepatocyte Review* Vol. Cap 2, Kluwer Academic Publishers.
4. Gomez-Lechon, M. J., Donato, M. T., Castell, J. V., et al. (2003) Human hepatocytes as a tool for studying toxicity and drug metabolism. *Curr Drug Metab* 4, 292–312.
5. Gomez-Lechon, M. J., Donato, M. T., Castell, J. V., et al. (2004) Human hepatocytes in primary culture: the choice to investigate drug metabolism in man. *Curr Drug Metab* 5, 443–462.
6. Vermeir, M., Annaert, P., Mamidi, R. N., et al. (2005) Cell-based models to study hepatic drug metabolism and enzyme induction in humans. *Expert Opin Drug Metab Toxicol* 1, 75–90.
7. LeCluyse, E. L. (2001) Human hepatocyte culture systems for the *in vitro* evaluation of cytochrome P450 expression and regulation. *Eur J Pharm Sci* 13, 343–368.
8. Mitry, R. R., Dhawan, A., Hughes, R. D., et al. (2004) One liver, three recipients: segment IV from split-liver procedures as a source of hepatocytes for cell transplantation. *Transplantation* 77, 1614–1616.
9. Richert, L., Alexandre, E., Lloyd, T., et al. (2004) Tissue collection, transport and isolation procedures required to optimize human hepatocyte isolation from waste liver surgical resections. A multilaboratory study. *Liver Int* 24, 371–378.
10. Laba, A., Ostrowska, A., Patrzalek, D., et al. (2005) Characterization of human hepatocytes isolated from non-transplantable livers. *Arch Immunol Ther Exp (Warsz)* 53, 442–453.
11. Hughes, R. D., Mitry, R. R., Dhawan, A., et al. (2006) Isolation of hepatocytes from livers from non-heart-beating donors for cell transplantation. *Liver Transpl* 12, 713–717.
12. LeCluyse, E. L., Alexandre, E., Hamilton, G. A., et al. (2005) Isolation and culture of primary human hepatocytes. *Methods Mol Biol* 290, 207–229.
13. Pichard, L., Raulet, E., Fabre, G., et al. (2006) Human hepatocyte culture. *Methods Mol Biol* 320, 283–293.
14. Bader, A. K. Haverich, A. (2000) in (Berry, M. N., Edwards, A. M., eds.), *The Hepatocyte Review*, pp. 97–116. Kluwer Academic Publuishers, London.
15. Serralta, A., Donato, M. T., Orbis, F., et al. (2003) Functionality of cultured human hepatocytes from elective samples, cadaveric grafts and hepatectomies. *Toxicol In Vitro* 17, 769–774.
16. Serralta, A., Donato, M. T., Martinez, A., et al. (2005) Influence of preservation solution on the isolation and culture of human hepatocytes from liver grafts. *Cell Transpl* 14, 837–843.
17. Donato, M. T., Serralta, A., Jimenez, N., et al. (2005) Liver grafts preserved in Celsior solution as source of hepatocytes for drug metabolism studies: comparison with surgical liver biopsies. *Drug Metab Dispos* 33, 108–114.
18. Muhlbacher, F., Langer, F., Mittermayer, C. (1999) Preservation solutions for transplantation. *Transpl Proc* 31, 2069–2070.
19. Janssen, H., Janssen, P. H., Broelsch, C. E. (2003) Celsior solution compared with University of Wisconsin solution (UW) and histidine-tryptophan-ketoglutarate solution (HTK) in the protection of human hepatocytes against ischemia-reperfusion injury. *Transpl Int* 16, 515–522.
20. Donato, M. T., Lahoz, A., Jimenez, N., et al. (2006) Potential impact of steatosis on cytochrome P450 enzymes of human hepatocytes isolated from fatty liver grafts. *Drug Metab Dispos* 34, 1556–1562.
21. Donato, M. T., Jimenez, N., Serralta, A., et al. (2007) Effects of steatosis on drug-metabolizing capability of primary human hepatocytes. *Toxicol In Vitro* 21, 271–276.
22. Turncliff, R. Z., Meier, P. J., Brouwer, K. L. (2004) Effect of dexamethasone treatment on the expression and function of transport proteins in sandwich-cultured rat hepatocytes. *Drug Metab Dispos* 32, 834–839.
23. Sidhu, J. S., Omiecinski, C. J. (1995) Modulation of xenobiotic-inducible cytochrome P450 gene expression by dexamethasone in primary rat hepatocytes. *Pharmacogenetics* 5, 24–36.
24. Runge, D., Runge, D. M., Jager, D., et al. (2000) Serum-free, long-term cultures of human hepatocytes: maintenance of cell morphology, transcription factors, and liver-specific functions. *Biochem Biophys Res Commun* 269, 46–53.
25. Katsura, N., Ikai, I., Mitaka, T., et al. (2002) Long-term culture of primary human hepatocytes with preservation of proliferative capacity and differentiated functions. *J Surg Res* 106, 115–123.

26. Donato, M. T., Ponsoda, X., O'Connor, E., et al. (2001) Role of endogenous nitric oxide in liver-specific functions and survival of cultured rat hepatocytes. *Xenobiotica* 31, 249–264.
27. Nakajima, H., Shimbara, N. (1996) Functional maintenance of hepatocytes on collagen gel cultured with simple serum-free medium containing sodium selenite. *Biochem Biophys Res Commun* 222, 664–668.
28. Fujita, R., Hui, T., Chelly, M., et al. (2005) The effect of antioxidants and a caspase inhibitor on cryopreserved rat hepatocytes. *Cell Transpl* 14, 391–396.
29. Pichard-Garcia, L., Gerbal-Chaloin, S., Ferrini, J. B., et al. (2002) Use of long-term cultures of human hepatocytes to study cytochrome P450 gene expression. *Methods Enzymol* 357, 311–321.
30. Chen, H. L., Wu, H. L., Fon, C. C., et al. (1998) Long-term culture of hepatocytes from human adults. *J Biomed Sci* 5, 435–440.
31. Du, Y., Chia, S. M., Han, R., et al. (2006) 3D hepatocyte monolayer on hybrid RGD/galactose substratum. *Biomaterials* 27, 5669–5680.
32. Richert, L., Liguori, M. J., Abadie, C., et al. (2006) Gene expression in human hepatocytes in suspension after isolation is similar to the liver of origin, is not affected by hepatocyte cold storage and cryopreservation, but is strongly changed after hepatocyte plating. *Drug Metab Dispos* 34, 870–879.
33. Dunn, J. C., Yarmush, M. L., Koebe, H. G., et al. (1989) Hepatocyte function and extracellular matrix geometry: long-term culture in a sandwich configuration. *Faseb J* 3, 174–177.
34. Lengyel, G., Veres, Z., Szabo, P., et al. (2005) Canalicular and sinusoidal disposition of bilirubin mono- and diglucuronides in sandwich-cultured human and rat primary hepatocytes. *Drug Metab Dispos* 33, 1355–1360.
35. Liu, X., LeCluyse, E. L., Brouwer, K. R., et al. (1999) Biliary excretion in primary rat hepatocytes cultured in a collagen-sandwich configuration. *Am J Physiol* 277, G12–21.
36. Koebe, H. G., Pahernik, S., Eyer, P., et al. (1994) Collagen gel immobilization: a useful cell culture technique for long-term metabolic studies on human hepatocytes. *Xenobiotica* 24, 95–107.
37. Talamini, M. A., Kappus, B., Hubbard, A. (1997) Repolarization of hepatocytes in culture. *Hepatology* 25, 167–172.
38. Selden, C., Shariat, A., McCloskey, P., et al. (1999) Three-dimensional *in vitro* cell culture leads to a marked upregulation of cell function in human hepatocyte cell lines – an important tool for the development of a bioartificial liver machine. *Ann N Y Acad Sci* 875, 353–363.
39. Clement, B., Guguen-Guillouzo, C., Campion, J. P., et al. (1984) Long-term co-cultures of adult human hepatocytes with rat liver epithelial cells: modulation of albumin secretion and accumulation of extracellular material. *Hepatology* 4, 373–380.
40. Tong, J. Z., Sarrazin, S., Cassio, D., et al. (1994) Application of spheroid culture to human hepatocytes and maintenance of their differentiation. *Biol Cell* 81, 77–81.
41. Khalil, M., Shariat-Panahi, A., Tootle, R., et al. (2001) Human hepatocyte cell lines proliferating as cohesive spheroid colonies in alginate markedly upregulate both synthetic and detoxificatory liver function. *J Hepatol* 34, 68–77.
42. Ijima, H., Nakazawa, K., Mizumoto, H., et al. (1998) Formation of a spherical multicellular aggregate (spheroid) of animal cells in the pores of polyurethane foam as a cell culture substratum and its application to a hybrid artificial liver. *J Biomater Sci Polym Ed* 9, 765–778.
43. Hamilton, G. A., Jolley, S. L., Gilbert, D., et al. (2001) Regulation of cell morphology and cytochrome P450 expression in human hepatocytes by extracellular matrix and cell-cell interactions. *Cell Tissue Res* 306, 85–99.
44. Gomez-Lechon, M. (1998) Primary culture of human hepatocytes, in (Doyle, A., Griffiths, J. B., eds.), *Cell and Tissue Culture Laboratory Procedures.*
45. Waxman, D. J., Lapenson, D. P., Aoyama, T. , et al. (1991) Steroid hormone hydroxylase specificities of eleven cDNA-expressed human cytochrome P450 s. *Arch Biochem Biophys* 290, 160–166.
46. Watford, M. (1991) The urea cycle: a two-compartment system. *Essays Biochem* 26, 49–58.
47. Rafoth, R. J., Onstad, G. R. (1975) Urea synthesis after oral protein ingestion in man. *J Clin Invest* 56, 1170–1174.
48. Donato, M. T., Castell, J. V. (2003) Strategies and molecular probes to investigate the role of cytochrome P450 in drug metabolism: focus on *in vitro* studies. *Clin Pharmacokinet* 42, 153–178.

Chapter 5

In Vitro Assays for Induction of Drug Metabolism

Brian G. Lake, Roger J. Price, Amanda M. Giddings and David G. Walters

Abstract

Hepatic microsomal cytochrome P450 (CYP) forms have a major role in the metabolism of drugs and other chemicals. Primary hepatocyte cultures from humans and experimental animals are a valuable in vitro system for studying the effects of chemicals on CYP forms. This chapter describes methods to evaluate CYP form induction in human and rat hepatocytes cultured in a 96-well plate format. The use of a 96-well plate format permits studies to be performed with relatively small numbers of hepatocytes and obviates the need to harvest cells and prepare subcellular fractions prior to the assay of enzyme activities. The induction of CYP1A and CYP3A forms in human and rat hepatocytes can be determined by measurement of 7-ethoxyresorufin *O*-deethylase and testosterone 6β-hydroxylase activities, respectively, whereas 7-benzyloxy-4-trifluoromethylcoumarin (BFC) *O*-debenzylase can be employed to assess both CYP1A and CYP2B form induction in rat hepatocytes. An assay for determining the protein content of hepatocytes cultured in a 96-well plate format is also described.

Key words: Cytochrome P450, 7-benzyloxy-4-trifluoromethylcoumarin *O*-debenzylase, enzyme induction, 7-ethoxyresorufin *O*-deethylase, human hepatocytes, rat hepatocytes, sulphorhodamine B protein assay, testosterone 6β-hydroxylase.

1. Introduction

Primary hepatocyte cultures are a valuable in vitro model system for studying many aspects of liver function and also for evaluating species differences in response. Mammalian hepatic cytochrome P450 (CYP) forms have a major role in the oxidative metabolism of drugs, food additives, pesticides, industrial chemicals, environmental contaminants and certain endogenous compounds *(1, 2)*. In the development of new therapeutic agents, it is important to ascertain whether the compound will be either an inhibitor or an inducer of hepatic CYP forms in order to exclude potential drug–drug interactions *(3, 4)*.

Anil Dhawan, Robin D. Hughes (eds.), *Hepatocyte Transplantation, vol. 481*

DOI 10.1007/978-1-59745-201-4_5 Springerprotocols.com

Many studies have demonstrated that primary hepatocyte cultures from humans and experimental animals (e.g. rodents) can be used to evaluate the effects of therapeutic agents and other chemicals on CYP forms *(5–12)*. To assess the induction of hepatic microsomal CYP forms, hepatocytes from humans and from species such as the rat can be cultured in conventional plates (e.g. 60 or 100 mm dishes), and at the end of the treatment period the hepatocytes harvested and microsomal fractions prepared by differential centrifugation. The induction of CYP forms can then be studied using the prepared microsomal fractions for either enzyme assays or Western immunoblotting for selected CYP forms. As an alternative, it is also possible to culture hepatocytes from humans and experimental animals in a 96-well plate format and assess enzyme induction by determining CYP-dependent enzyme activities in intact hepatocytes *(13–17)*. CYP form induction may also be assessed by assaying CYP mRNA levels in cells cultured in a 96-well plate format and in other formats.

This chapter describes three CYP-dependent enzyme assays that can be performed in human and rat hepatocytes cultured in a 96-well plate format and an assay for hepatocyte protein content that can be used to normalise the results of the CYP-dependent enzyme activity measurements. The use of 7-ethoxyresorufin *O*-deethylase activity as a marker for induction of CYP1A forms in human and rat hepatocytes cultured in a 96-well plate format has been previously described by Castell and co-workers *(13, 14)*. Many studies have demonstrated that testosterone 6β-hydroxylase is a specific marker for CYP3A forms in both human and rodent liver and this activity may also be used as a marker for CYP3A form induction in cultured hepatocytes *(2, 6, 9, 10, 12)*. Studies with rat hepatocytes have demonstrated that 7-benzyloxy-4-trifluoromethylcoumarin (BFC) *O*-debenzylase activity is a good marker for the induction of both CYP1A and CYP2B forms *(16)*. In human hepatocytes, this enzyme activity may be a marker for CYP1A and possibly also CYP3A forms. When using intact cells, rather than subcellular fractions, for CYP enzyme activity determinations, attention needs to be paid to the possible phase II metabolism of the CYP substrates employed. With the 7-ethoxyresorufin *O*-deethylase assay, the resorufin product can be a substrate for cytosolic quinone reductase and is also conjugated with D-glucuronic acid and sulphate *(13, 14)*. The need for enzymatic deconjugation also applies to the assay of BFC O-debenzylase activity *(16)*, whereas no enzymatic deconjugation is required for the testosterone 6β-hydroxylase assay *(8, 14)*. This chapter also describes the sulphorhodamine B (SRB) protein assay for hepatocyte protein content in a 96-well plate format. This assay was developed by Boyd and co-workers for use in anti-cancer drug screening in cell lines *(18, 19)* and represents a convenient assay to normalise CYP-dependent enzyme activities in hepatocytes cultured in a 96-well plate format. Finally, while this chapter focuses on the induction of CYP forms, methods for assessing the inhibition of CYP forms in cultured hepatocytes have been described elsewhere *(6)*.

2. Materials

2.1. Reference Items for Hepatocyte Culture

1. Dimethyl sulphoxide (DMSO). A high-purity grade (e.g. ≥ 99.9%) should be used.
2. 0.2, 2 and 20 mM β-Naphthoflavone (BNF; Sigma-Aldrich Chemical Company, Poole, Dorset, UK) in DMSO. Store in aliquots at –20°C, thaw only once.
3. 2 and 10 mM Rifampicin (rifampin; RIF; Sigma-Aldrich) in DMSO. Store in aliquots at –20°C, thaw only once.
4. 2 and 20 mM Pregnenolone 16α-carbonitrile (PCN; Sigma-Aldrich) in DMSO. Store in aliquots at –20°C, thaw only once.
5. 20 mM Sodium phenobarbitone (phenobarbital; NaPB; Sigma-Aldrich). This reference item is dissolved directly in tissue culture medium and then diluted with tissue culture medium to final concentrations of 200 and 500 µM. Prepare immediately before use.

2.2. For 7-Ethoxyresorufin O-Deethylase Assay

1. RPMI 1640 (phenol red free) medium (Invitrogen Ltd., Paisley, Scotland, UK).
2. 2 mM 7-ethoxyresorufin (Sigma-Aldrich) in DMSO. Store in aliquots at –20°C, thaw only once.
3. 20 mM Dicumarol (Sigma-Aldrich) in DMSO. Store in aliquots at –20°C, thaw only once.
4. 0.5 M Sodium acetate buffer, pH 5.0. Store at room temperature, discard after 12 weeks.
5. β-Glucuronidase/sulphatase solution. Dilute combined β-glucuronidase/arylsulphatase preparation (Catalogue no. 10127060001, from *Helix pomatia*) obtained from Roche Diagnostics Ltd. (Lewes, Sussex, UK) 1:100 with deionised water. Prepare immediately before use.
6. 0.25 M Tris (i.e. 30.275 g/1000ml) in 60% (v/v) acetonitrile (ACN). Store at room temperature, discard after 12 weeks.
7. 2 mM Resorufin (Sigma-Aldrich) in ethanol. Store in aliquots at –20°C, thaw only once.

2.3. For BFC O-Debenzylase Assay

1. RPMI 1640 (phenol red free) medium (Invitrogen Ltd.).
2. 12.5 mM BFC (Sigma-Aldrich) in DMSO. Store in aliquots at –20°C, thaw only once.
3. 0.5 M Sodium acetate buffer, pH 5.0. Store at room temperature, discard after 12 weeks.
4. β-Glucuronidase/sulphatase solution. Dilute combined β-glucuronidase/arylsulphatase preparation (Catalogue No. 10127060001, from *Helix pomatia*) obtained from Roche Diagnostics Ltd. (Lewes, Sussex, UK) 1:100 with deionised water. Prepare immediately before use.
5. 0.25 M Tris (i.e. 30.275 g/1000 ml deionised water) in 60% (v/v) ACN. Store at room temperature, discard after 12 weeks.
6. 0.6667 mM 7-Hydroxy-4-trifluoromethylcoumarin (HFC; Sigma-Aldrich) in DMSO. Store in aliquots at –20°C, thaw only once.

2.4. For Testosterone 6β-Hydroxylase Assay

1. RPMI 1640 (phenol red free) medium (Invitrogen Ltd.).
2. [4-^{14}C]Testosterone (e.g. specific activity around 54 mCi/mmol, CFA129, from GE Healthcare UK Ltd., Little Chalfont, Bucks, UK) and unlabelled testosterone and 6β-hydroxytestosterone (Sigma-Aldrich).
3. High-performance liquid chromatography (HPLC) grade ACN and methanol.

2.5. For Hepatocyte Protein Assay

1. 10% (w/v) Trichloroacetic acid (TCA). Store at room temperature, discard after 12 weeks.
2. 1% (v/v) Glacial acetic acid. Store at room temperature, discard after 12 weeks.
3. 0.4% (w/v) SRB (Sigma-Aldrich) in 1% (v/v) glacial acetic acid. Prepare immediately before use.
4. 10 mM Tris (i.e. 1.211 g/1000 ml deionised water). Store at 4°C, discard after 12 weeks.

3. Methods

3.1. Treatment with Test Compounds and Reference Items

1. The CYP form activities described in this chapter are suitable for use with primary human and rat hepatocytes cultured in a 96-well plate format, employing a seeding density of around 30,000 viable cells/well. The use of a sandwich culture technique (e.g. use of plates coated with a suitable extracellular matrix such as collagen, fibronectin

or Matrigel® and the attached hepatocytes then overlaid with extracellular matrix) is recommended *(5, 7, 9, 10)*. Human and rat hepatocytes are normally cultured in control medium for 1–3 days before being treated with CYP form inducers *(5,6,7,10)*. To study the induction of CYP forms, primary hepatocyte cultures are treated with the test compounds (i.e. the compounds under investigation) and reference items (see below) for a suitable period (e.g. 2 or 3 days). Normally the culture medium is changed at 24 h intervals and replaced with fresh medium containing the test compounds and reference items. Test compounds and reference items may be added to the culture medium in DMSO (*see* **Note 1**).

2. When employing 96-well plates, replicates (*see* **Note 2**) are normally performed for both cells cultured in control medium and for cells treated with the test compounds and reference items. For the 7-ethoxyresorufin *O*-deethylase and BFC *O*-debenzylase fluorescent assays, up to 12 wells/plate should be controls (i.e. hepatocytes cultured in control medium containing the DMSO solvent) and up to 6 wells/plate for each concentration of each test compound and reference item. With the radiometric testosterone 6β-hydroxylase assay, it may be necessary to pool two or three wells for each control and treatment in order to provide a sufficient volume of incubation medium for HPLC analysis.
3. For all assays, suitable blanks should be run in parallel with the treatment of the hepatocyte preparations. These consist of incubations in 96-well plates containing the overlay (e.g. collagen or Matrigel®) and control medium but no hepatocytes. For the two fluorescent assays, eight blank wells are normally sufficient, whereas for the radiometric assay up to four wells or four pools of two or more wells may be required.
4. To assess the functional viability of human and rat hepatocyte preparations for CYP form induction studies, the use of reference items is recommended. Suitable reference item concentrations (*see* **Note 3**) are as follows:

(a) For CYP1A form induction in human hepatocytes use 2 and 10 μM BNF and for rat hepatocytes use 0.2 and 2 μM BNF.

(b) For CYP2B form induction in rat hepatocytes use 200 and 500 μM NaPB.

(c) For CYP3A form induction in human hepatocytes use 2 and 10 μM RIF. Studies may also be conducted with 200 and 500 μM NaPB.

(d) For CYP3A form induction in rat hepatocytes use 2 and 20 μM PCN.

3.2. Assay of 7-Ethoxyresorufin O-Deethylase Activity

1. Prepare sufficient 7-ethoxyresorufin substrate solution (added at 100 μl/well) for all wells and plates to be assayed, by thawing aliquots stored at –20°C of 2 mM 7-ethoxyresorufin in DMSO and 20 mM dicumarol in DMSO. Add 4 μl/ml 2 mM 7-ethoxyresorufin and 0.5 μl/ml 20 mM dicumarol in DMSO per millilitre of RPMI 1640 (phenol red free) medium at 37°C. Mix the substrate solution (final concentrations 7-ethoxyresorufin 8 μM and dicumarol 10 μM) with a vortex mixer and return to the incubator.
2. At the end of the treatment period with the test compounds and the reference items, the medium is removed and the cells washed with 200 μl/well of RPMI 1640 (phenol red free) medium at 37°C. Return the plates to the incubator.
3. Remove the RPMI 1640 (phenol red free) wash medium from each plate and quickly add 100 μl/well of the 8 μM 7-ethoxyresorufin/10 μM dicumarol substrate solution to each well and mix the plates for 5 s on a gyratory shaker.
4. Return the plates to the tissue culture incubator and incubate for a suitable period (e.g. 30 and 20 min for human and rat hepatocytes, respectively) at 37°C (*see* **Note 4**).
5. At the end of the incubation period, mix the plates on a gyratory shaker for 5 s and remove a 75 μl aliquot of the medium from each well into a "V"-bottomed 96-well plate and store at –80°C prior to analysis.
6. Thaw the "V"-bottomed 96-well plates and add 10 μl/well 0.5 M sodium acetate buffer pH 5.0 and 15 μl/well of the β-glucuronidase/sulphatase solution (*see* **Section 2.2**) to all wells, mix the plates for 5 s on a gyratory shaker and incubate for 2 h at 37°C.
7. Prepare a 2 μM resorufin standard by thawing an aliquot of 2 mM resorufin in DMSO and diluting 10 μl to a final volume of 10 ml with RPMI 1640 (phenol red free) medium. Set up a standard curve by adding 0 (blank), 5, 10, 15, 20, 25, 30, 40 and 50 μl aliquots of the 2 μM resorufin standard to a "V"-bottomed 96-well plate (for the standard curve use eight replicate wells for each resorufin concentration) and add 25–75 μl/well of RPMI 1640 (phenol red free) medium so that each well has a final volume of 75 μl. Add 10 μl/well 0.5 M sodium acetate buffer pH 5.0 and 15 μl/well of the β-glucuronidase/sulphatase solution (*see* **Section 2.2**) to all wells, mix the plates for 5 s on a gyratory shaker and incubate for 2 h at 37°C.
8. At the end of the incubation period, add 100 μl of 0.25 M Tris in 60% (v/v) ACN to all wells and mix the plates on a gyratory shaker for 15 s. Transfer 150 μl from each well into a flat-bottomed white polystyrene 96-well plate. Set up a fluorescence spectrophotometer with a 96-well plate reader and determine the fluorescence of each well at excitation and

emission wavelengths of 535 and 582 nm, respectively (*see* **Note 5**).

9. For the resorufin standard curve, subtract the mean fluorescence of the blank wells (no resorufin standard) and plot fluorescence units against picomole of resorufin added (in the 150 μl sample analysed, the resorufin standards range from 7.5 to 75 pmol).
10. For the hepatocyte samples, subtract the mean fluorescence of the blank wells (i.e. the wells containing no hepatocytes) from the test wells and using the standard curve (see above) determine the picomole resorufin formed per well. By allowing for the incubation time, the results are expressed either as picomole resorufin formed per minute per number of cells per well or with the hepatocyte protein content of each well (*see* **Section 3.5**) as picomole resorufin formed per minute per microgram hepatocyte protein.

3.3. Assay of BFC O-Debenzylase Activity

1. Prepare sufficient BFC substrate solution (added at 100 μl/well) for all wells and plates to be assayed, by thawing aliquots stored at –20°C of 12.5 mM BFC. Add 4 μl/ml 12.5 mM BFC per millilitre of RPMI 1640 (phenol red free) medium at 37°C. Mix the substrate solution (final BFC concentration 50 μM) with a vortex mixer and return to the incubator.
2. At the end of the treatment period with the test compounds and the reference items, the medium is removed and the cells washed with 200 μl/well of RPMI 1640 (phenol red free) medium at 37°C. Return the plates to the incubator.
3. Remove the RPMI 1640 (phenol red free) wash medium from each plate and quickly add 100 μl/well of the 50 μM BFC substrate solution to each well and mix the plates for 5 s on a gyratory shaker.
4. Return the plates to the tissue culture incubator and incubate for a suitable period (e.g. 20 min for rat hepatocytes) at 37°C (*see* **Note 4**).
5. At the end of the incubation period, mix the plates on a gyratory shaker for 5 s and remove a 75 μl aliquot of the medium from each well into a "V"-bottomed 96-well plate and store at –80°C prior to analysis.
6. Thaw the "V"-bottomed 96-well plates and add 10 μl/well 0.5 M sodium acetate buffer pH 5.0 and 15 μl/well of the β-glucuronidase/sulphatase solution (*see* **Section 2.3**) to all wells, mix the plates for 5 s on a gyratory shaker and incubate for 2 h at 37°C.
7. Prepare a 6.667 μM HFC standard by thawing an aliquot of 0.6667 mM HFC in DMSO and diluting 100 μl to a final volume of 10 ml with RPMI 1640 (phenol red free) medium.

Set up a standard curve by adding 0 (blank), 5, 10, 15, 20, 25, 30, 40 and 50 μl aliquots of the 6.667 μM HFC standard to a "V"-bottomed 96-well plate (for the standard curve use 8 replicate wells for each HFC concentration) and add 25–75 μl/well of RPMI 1640 (phenol red free) medium so that each well has a final volume of 75 μl. Add 10 μl/well 0.5 M sodium acetate buffer pH 5.0 and 15 μl/well of the β-glucuronidase/sulphatase solution (*see* **Section 2.3**) to all wells, mix the plates for 5 s on a gyratory shaker and incubate for 2 h at 37°C.

8. At the end of the incubation period, add 100 μl of 0.25 M Tris in 60% (v/v) ACN to all wells and mix the plates on a gyratory shaker for 15 s. Transfer 150 μl from each well into a flat-bottomed white polystyrene 96-well plate. Set up a fluorescence spectrophotometer with a 96-well plate reader and determine the fluorescence of each well at excitation and emission wavelengths of 410 and 510 nm, respectively (*see* **Note 5**).
9. For the HFC standard curve, subtract the mean fluorescence of the blank wells (no HFC standard) and plot fluorescence units against picomole of HFC added (in the 150 μl sample analysed the HFC standards range from 25 to 250 pmol).
10. For the hepatocyte samples, subtract the mean fluorescence of the blank wells (i.e. the wells containing no hepatocytes) from the test wells and using the standard curve (see above) determine the picomole HFC formed per well. By allowing for the incubation time, the results are expressed either as picomole HFC formed per minute per number of cells per well or with the hepatocyte protein content of each well (*see* **Section 3.5**) as picomole HFC formed per minute per milligram hepatocyte protein.

3.4. Assay of Testosterone 6β-Hydroxylase Activity

1. Prepare sufficient 250 μM [4-^{14}C]testosterone substrate solution to add at 100 μl/well with each well receiving 0.4 μCi radioactivity. For example, 10 ml of substrate solution will contain 2.5 μmol testosterone and 40 μCi radioactivity. Add 40 μCi of stock [4-^{14}C]testosterone to a tapered glass tube and remove the solvent with a stream of nitrogen. Then add 10 μl of DMSO containing unlabelled testosterone so that the tube contains a total of 2.5 μmol of labelled and unlabelled testosterone. For a specific activity of 54 mCi/mmol, the unlabelled testosterone substrate solution will be 175.93 mM. Vortex the tube contents and transfer the DMSO solvent to 10 ml of RPMI 1640 (phenol red free) medium at 37°C and mix well with a vortex mixer. Add 10 μl of DMSO to the tapered glass tube, vortex the tube contents and transfer to the RPMI 1640 (phenol red free) medium at 37°C. Repeat with two further 10 μl and one 5 μl washes of DMSO. Mix the 250 μM

[4-^{14}C]testosterone substrate with a vortex mixer and return to the incubator.

2. At the end of the treatment period with the test compounds and the reference items, the medium is removed and the cells washed with 200 μl/well of RPMI 1640 (phenol red free) medium at 37°C. Return the plates to the incubator.
3. Remove the RPMI 1640 (phenol red free) wash medium from each plate and quickly add 100 μl/well of the 250 μM [4-^{14}C]testosterone substrate solution to each well and mix the plates for 5 s on a gyratory shaker.
4. Return the plates to the tissue culture incubator and incubate at 37°C for a suitable period (e.g. 30 and 20 min for human and rat hepatocytes, respectively) at 37°C (*see* **Note 4**).
5. At the end of the incubation period, mix the plates on a gyratory shaker for 5 s and remove the medium from all wells into Eppendorf tubes, pooling wells as required (*see* **Section 3.1**). Store the tubes at –80°C prior to analysis.
6. Thaw the samples and analyse aliquots by HPLC, employing a 150×4.6 mm column of Supelcosil-5 LC-18 (Sigma-Aldrich) protected by a 20×4.6 mm column of Supelcosil-5 LC-18 and mobile phases of ACN (A), ultrapure water (B), methanol (C) and 10% (v/v) acetic acid in ultrapure water (D). Elution is achieved at a flow rate of 2 ml/min starting with 12% A, 73% B, 10% C and 5% D for 10 min, changing to 12% A, 67% B, 16% C and 5% D over 14.2 min, changing to 14% A, 81% C and 5% D over 1 min, holding at 14% A, 81% C and 5% D for 4 min, changing to 12% A, 73% B, 10% C and 5% D over 0.8 min, holding at 12% A, 73% B, 10% C and 5% D for 4 min and equilibrating at 12% A, 73% B, 10% C and 5% D for 4 min before the next injection. Retention times of testosterone and 6β-hydroxytestosterone are approximately 18 and 14 min, respectively. Formation of 6β-hydroxytestosterone is quantified by radiometric detection (*see* **Note 6**).
7. The amount of 6β-hydroxytestosterone formed in the sample less any material present in the blank (no hepatocytes) incubations is determined as a percentage of the substrate added (25 nmol per well). By allowing for the incubation time, the results are expressed either as picomole 6β-hydroxytestosterone formed per minute per number of cells per well or with the hepatocyte protein content of each well (*see* **Section 3.5**) as picomole 6β-hydroxytestosterone formed per minute per milligram hepatocyte protein.

3.5. Assay of Hepatocyte Protein Content

1. At the end of the incubations with the CYP substrates, all remaining medium is removed and 100 μl of 10% (w/v) TCA added to each well and the plates stored at 4°C for 30 min.

2. Remove the TCA solution from the plates by inverting and shaking the plates. Wash all wells of each plate four times with deionised water, inverting the plates and tapping on a paper towel between each wash. Allow plates to air dry and store at 4°C prior to analysis.
3. Add 50 μl/well of 0.4% (w/v) SRB in 1% (v/v) glacial acetic acid and leave the plates for 30 min at room temperature.
4. Remove the unbound dye from the plates by inverting and shaking the plates. Rapidly wash the wells of each plate four times with 1% (v/v) glacial acetic acid, inverting the plates and tapping on paper towel between each wash. Do not allow the acetic acid wash to remain on the cells for longer than a few seconds. After the final wash allow the plates to air dry.
5. Add 200 μl/well of 10 mM Tris and place the plates on a gyratory shaker for 5 min.
6. Set up a 96-well plate reader and determine the absorbance of each well at 490 nm, employing 630 nm as a reference wavelength. Subtract the mean of the blank wells (i.e. the wells containing no hepatocytes) from each of the test wells.
7. For rat hepatocytes, the absorbance values can be multiplied by 102 to convert SRB assay absorbance units into microgram hepatocyte protein per well (*see* **Note** 7).

4. Notes

1. DMSO is a good solvent for many chemicals and at low concentrations it is not cytotoxic to hepatocytes. However, DMSO is a known inducer of CYP3A4 in cultured human hepatocytes *(9, 10)* and hence final medium concentrations should be $\leq$ 0.1% (v/v).
2. The number of replicates required is dependent on a number of factors, including the precision required and the magnitude of the effect of the test compounds.
3. The concentrations of the reference items quoted for human and rat hepatocytes are a guide only and are dependent on the experimental conditions including the treatment period. For a given set of experimental conditions, it is recommended that a range of concentrations of each reference item is examined to identify suitable concentrations for subsequent experiments. For CYP3A induction, RIF (not PCN) should be employed as a reference item for human hepatocytes, whereas PCN (not RIF) should be selected as a reference item for rat hepatocytes *(6)*.
4. The incubation times quoted for human and rat hepatocytes are a guide only. Enzyme activity is dependent on the plating

density and the time period that the cells are cultured before treatment is commenced and the period of treatment with the test compounds. It is recommended that the linearity of each assay with time of incubation is established for a given set of experimental conditions.

5. The wavelengths cited are for use with a fluorescence spectrophotometer with a 96-well plate attachment. For filter instruments, select filters with the nearest available wavelengths to those cited above.
6. Many HPLC methods are available for the separation of testosterone and its metabolites *(8, 14)*. As an alternative, the analysis of testosterone 6β-hydroxylase activity in hepatocytes cultured in a 96-well plate format can also be determined with unlabelled substrate, the product being analysed by liquid chromatography–mass spectrometry–mass spectrometry *(20)*.
7. This factor reported for rat hepatocytes *(16)* may also be applied to human hepatocytes. It is also possible to utilise the Lowry assay to determine the protein content of hepatocytes cultured in a 96-well plate format *(15)*.

References

1. Lewis, D. F. V. (2001) *Guide to Cytochromes P450: Structure and Function,* Taylor and Francis, London.
2. Parkinson, A. (2001) Biotransformation of xenobiotics, in (Klaassen, C. D., ed.), *Casarett and Doull's Toxicology: The Basic Science of Poisons*, 6th edn, pp. 133–224. McGraw Hill, New York.
3. Pelkonen, O., Mäenpää, J., Taavitsainen, P., et al. (1998). Inhibition and induction of human cytochrome P450 (CYP) enzymes. *Xenobiotica* 28, 1203–1253.
4. Lin, J. H., Lu, A. Y. H. (1998) Inhibition and induction of cytochrome P450 and the clinical implications. *Clin. Pharmacokinet* 35, 361–390.
5. Sidhu, J. S., Farin, F. M., Omiecinski, C. J. (1993) Influence of extracellular matrix overlay on phenobarbital-mediated induction of CYP2B1, 2B2 and 3A1 genes in primary adult rat hepatocyte culture. *Arch. Biochem Biophys* 301, 103–113.
6. Maurel, P. (1996) The use of adult human hepatocytes in primary culture and other in vitro systems to investigate drug metabolism in man. *Adv Drug Deliv Rev* 22, 105–132.
7. Coecke, S., Rogiers, V., Bayliss, M., et al. (1999) The use of long-term hepatocyte cultures for detecting induction of drug metabolising enzymes: the current status. ECVAM Hepatocytes and Metabolically Competent Systems Task Force Report 1. *ATLA* 27, 579–638.
8. Kostrubsky, V. E., Ramachandran, V., Venkataramanan, R., et al. (1999) The use of human hepatocytes to study the induction of cytochrome P-450. *Drug Metab Dispos.* 27, 887–894.
9. LeCluyse, E. L. (2001) Human hepatocyte culture systems for the in vitro evaluation of cytochrome P450 expression and regulation. *Eur J Pharm Sci* 13, 343–368.
10. LeCluyse, E., Madan, A., Hamilton, G., et al. (2000) Expression and regulation of cytochrome P450 enzymes in primary cultures of human hepatocytes. *J Biochem Mol Toxicol* 14, 177–188.
11. Gerbal-Chaloin, S., Pascussi, J.-M., Pichard-Garcia, L., et al. (2001) Induction of CYP2C genes in human hepatocytes in primary culture. *Drug Metab Dispos* 29, 242–251.
12. Parkinson, A., Mudra, D. R., Johnson, C., et al. (2004) The effects of gender, age, ethnicity,

and liver cirrhosis on cytochrome *P*450 enzyme activity in human liver microsomes and inducibility in cultured human hepatocytes. *Toxicol Appl Pharmacol* 199, 193–209.

13. Donato, M. T., Gómez-Lechón, M. J., Castell, J. V. (1993) A microassay for measuring cytochrome P450IA1 and P450IIB1 activities in intact human and rat hepatocytes cultured on 96-well plates. *Anal Biochem* 213, 29–33.
14. Gómez-Lechón, M. J., Donato, T., Ponsoda, X., et al. (1997) Isolation, culture and use of human hepatocytes in drug research, in (Castell, J.V., and Gómez-Lechón, M.J., eds.), *In Vitro Methods in Pharmaceutical Research*, pp.129–153. Academic Press, London.
15. Donato, M. T., Castell, J. V., Gómez-Lechón, M. J. (1998) The coumarin 7-hydroxylation microassay in living cells in culture. *ATLA* 26, 213–223.
16. Price, R. J., Surry, D., Renwick, A. B., et al. (2000) CYP isoform induction screening in 96-well plates: use of 7-benzyloxy-4-trifluoromethylcoumarin as a substrate for studies with rat hepatocytes. *Xenobiotica* 30, 781–795.
17. Nicoll-Griffith, D. A., Chauret, N., Houle, R., et al. (2004) Use of a benzyloxy-substituted lactone cyclooxygenase-2 inhibitor as a selective fluorescent probe for CYP3A activity in primary cultured rat and human hepatocytes. *Drug Metab Dispos* 32, 1509–1515.
18. Skehan, P., Storeng, R., Scudiero, D., et al. (1990) New colorimetric assay for anticancer-drug screening. *J Natl Cancer Inst* 82, 1107–1112.
19. Rubenstein, L. V., Shoemaker, R. H., Paull, K. D., et al. (1990) Comparison of *in vitro* anticancer-drug-screening data generated with a tetrazolium assay versus a protein assay against a diverse panel of human tumor cell lines. *J Natl Cancer Inst* 82, 1113–1118.
20. Burczynski, M. E., McMillian, M., Parker, J. B., et al. (2001) Cytochrome P450 induction in rat hepatocytes assessed by quantitative real-time reverse-transcription polymerase chain reaction and the RNA invasive cleavage assay. *Drug Metab Dispos* 29, 1243–1250.

Chapter 6

Hepatocyte Apoptosis

Mustapha Najimi, Françoise Smets, and Etienne Sokal

Abstract

Apoptosis has been documented as a frequent hurdle phenomenon that occurs in human hepatocytes during isolation, storage, infusion and after engraftment within the recipient liver parenchyma. Apoptosis is an active form of cell death that involves programmed cellular machineries leading to a progressive self-destruction of the cell. In contrary to necrosis, it can affect individual cells within a cell population. It is characterized by chronological alteration of intracellular biochemical signaling pathways followed by cellular morphological changes, DNA fragmentation, perturbation of mitochondrial membrane function and changes in the plasma membrane. These cellular alterations can be analyzed using different methodologies on adherent, suspended and in situ engrafted hepatocytes. This chapter presents a brief overview of these techniques and provides methodology for the evaluation of hepatocyte apoptosis at the structural and biochemical levels.

Key words: Apoptosis, hepatocytes, DNA fragmentation, electrophoresis, agarose gel, caspase activation, immunoblotting, spectrophotometer, cell death, flow cytometry, mitochondria, TUNEL, immunohistochemistry, histology, cytology.

1. Introduction

Massive cell loss remains a limiting factor for the long-term success and durability of liver cell transplantation. It is basically the consequence of cell detachment from the extracellular matrix during isolation and cryopreservation/thawing steps *(1, 2)*. Hence, the quality of hepatocytes suspension dedicated to transplantation is investigated before infusion for a rapid evaluation of specific parameters as for instance cell viability and metabolic activity (*see* **Note 1**). Classical assays that are widely used for that purpose are trypan blue dye exclusion test, lactate dehydrogenase leakage and 3-(4,5-dimethyl-2-thiazolyl)-2,5-diphenyl-2H tetrazolium bromide assays. Intracellular ATP levels, because

Anil Dhawan, Robin D. Hughes (eds.), *Hepatocyte Transplantation, vol. 481*

DOI 10.1007/978-1-59745-201-4_6 Springerprotocols.com

hepatocytes are highly metabolic cells, could also be analyzed to investigate the quality of cell suspension. These assays are basically used because they are relatively quite simple and need equipment that could be found in all laboratories. However, they cannot specifically inform about the presence of apoptosis but remain only informative regarding the presence and the level of cell death in the analyzed cell population. With respect to apoptosis, analyses conducted on attached or suspended hepatocytes, after isolation or cryopreservation/thawing, may combine complementary rapid and slow techniques. Nevertheless, evaluation of cell morphology and nuclear staining remain the quickest and the gold standard assays for apoptosis studies and to distinguish this cell death phenomenon from necrosis.

Apoptotic death is the result of a succession of intracellular events that occur in response to several signals. It can be detected at its early reversible or late irreversible stages, thanks to the characterization of their mechanistic pathways. With respect to liver cell transplantation, apoptosis has to be evaluated after hepatocyte isolation, cryopreservation/thawing and infusion within the recipient liver even if the classical tests detect any alteration of hepatocyte viability.

2. Materials

2.1. Nuclear Staining

- Microscope coverslips of 12 mm diameter (VWR, Leuven, Belgium).
- Hoechst 33258 (Invitrogen, Merelbeke, Belgium) and 4'-6-diamino-2-phenylindole dihydrochloride (Sigma, Bornem, Belgium) are sensitive to light and can be dissolved in deionized water or DMSO (at 10 and 5 mg/mL, respectively).
- Propidium iodide (PI) (Sigma) can be dissolved in deionized water at 1 mg/mL and stored up to 6 months at 4°C in the dark.
- Successive dilutions of the stock solutions of these dyes can be performed in phosphate-buffered saline (PBS).
- For long-term storage, aliquots of stock solution of these dyes can be stored at $\leq$ –20°C or at 4°C for short-term use (the stock solutions can be stable for up to 6 months).
- All these dyes are mutagenic and must be carefully handled.
- Paraformaldehyde (Sigma): Prepare a 4% (w/v) fresh solution in PBS. The solution needs to be carefully heated for dissolution (using a stirring hot plate in a fume hood). The prepared solution must be cooled at room temperature before use.

2.2. Agarose Gel Electrophoresis

- Hepatocyte suspension at 3–5 million cells/mL in Williams' medium (Invitrogen) supplemented with 10% fetal calf serum (AE Scientific, Marcq, Belgium), 25 ng/mL epidermal growth factor (Peprotech, London, UK), 10 μg/mL insulin (Eli Lilly, Belgium), 1 μM dexamethasone (Sigma).
- Phosphate-citrate buffer, pH 7.8: 192 parts of 0.2 M Na_2HPO_4 and 8 parts of 0.1 M citric acid (pH 7.8).
- TBE buffer: prepare 10× stock with 89 mM Tris base, 89 mM boric acid, 0.5 M EDTA, pH 8 and store at room temperature. Dilute 1:10 before use.
- Ethidium bromide (EB): dissolve 50 mg in 100 mL H_2O and dilute 1:1000 before use. This is a mutagenic reagent, which must be carefully handled.
- Electrophoresis grade agarose (Invitrogen): dissolve 1% in 1× TBE by heating until melted before adding EB.
- DNA molecular weight markers (Fermentas, St.Leon-Rot, Germany).

2.3. TUNEL Assay

- Proteinase K 20 μg/mL (Roche, Brussels, Belgium).
- 2% H_2O_2 (Sigma) is used in non-fluorescent **T**erminal deoxyribonucleotidyl transferase (TdT) mediated d**UTP** **N**ick **E**nd **L**abeling (TUNEL) assays to inactivate the endogenous peroxidase.
- Triton X-100 at 0.1% in PBS.
- In Situ Cell Death Detection Kits (Roche).

2.4. Flow Cytometry

- Cell suspension at a concentration of 500–1000 cells/μL PBS.
- BD Bioscience binding Buffer.
- FITC-labeled Annexin V (BD Bioscience): dilute at 1 mg/mL in binding buffer.
- PI (Sigma) to dilute at 10 mg/mL in binding buffer.

2.5. Determination of Mitochondrial Membrane Potential

- Rhodamine 123: dilute at 1 mg/mL in ethanol and store at –20°C in the dark. Handle with care.

2.6. Analysis of Cytoplasmic Cell Compartment

- Hepatocytes at a concentration of 1.5×10^7 cells/mL Williams' medium.
- Permeabilization medium: 0.25 M sucrose, 3 mM EDTA-Na^+, 20 mM MOPS and 110 μg/mL digitonin, pH 7.4.
- Lysis buffer: 150 mM NaCl, 50 mM Tris–HCl pH 7.5, 0.5% deoxycholate, 1% NP-40 and 0.1% SDS.
- TBS buffer: 50 mM Tris pH 8.1, 150 mM NaCl.
- Blocking buffer: TBS containing 5% non-fat dry milk.
- Antibody dilution buffer: TBS containing 0.05% Tween 20.

- Purified mouse anti-cytochrome C monoclonal antibody (Becton Dickinson, clone 7H8.2C12)
- Secondary antibody: Anti-mouse IgG conjugated to horseradish peroxidase (GE Healthcare, Diegem, Belgium).
- Enhanced chemiluminescent (ECL) reagents (PerkinElmer, Zaventem, Belgium).

2.7. Caspase Activity

- One million hepatocytes suspended in Williams' medium or PBS.
- 96-Well flat-bottomed plates (Greiner Bio One, Wemmel, Belgium).
- Caspase-3/CPP32 and Caspase-8/FLICE fluorometric assay kits (Gentaur, Brussels, Belgium) containing cell lysis buffer, 2× reaction buffer, the corresponding labeled substrate and dithiothreitol 1 M.

2.8. Transmission Electron Microscopy

- Hepatocytes suspended at a concentration of $1–5 \times 10^6$ cells in Williams' medium.
- 2.5% EM grade glyceraldehyde (Agar Scientific) buffered in 0.1 M sodium cacodylate
- 1% osmium tetroxide (Agar Scientific).
- Epoxy Embedding Medium (Fluka Chemie, Buchs, Switzerland).
- Lead citrate, practically insoluble in water, is soluble at high concentrations in basic solutions. The staining solution, stored in glass or polyethylene bottles, is stable up to 6 months. If long term stored, centrifuge the solution before use.
- Zeiss EM109 transmission electron microscope (Carl Zeiss Inc., Oberkochem, Germany).

3. Methods

3.1. Morphological Evaluation

3.1.1. Nuclear Staining

Amongst the well-described features of apoptotic cells, nuclear changes such as chromatin condensation and nuclear fragmentation are the result of DNA cleavage by endogenous nucleases into oligonucleosomal fragments. This leads to the formation of dense and crescent-shaped chromatin aggregates. Other events chronologically linked are also documented, such as nuclear shrinkage and the formation of dense and granular nuclear particles termed apoptotic bodies. Such alterations can easily be revealed using specific nuclear dyes and observed by microscopy. The assay is simple, rapid and has the advantage to analyze a large number of cells for accurate quantification.

3.1.1.1. Fluorescent Dyes

The use of fluorescent dyes is very useful for the evaluation of apoptosis on cultured or cytocentrifuged hepatocytes but needs high-technology materials, such as fluorescence microscopy, for the observation and evaluation of the staining.

3.1.1.1.1. Hoechst 33258 Hoechst 33258 (bisbenzimide) is a cell-permeant nucleic acid stain that is taken by all cells and emits blue fluorescence after UV excitation (excitation/emission maxima of 360/450 nm, respectively). The reagent can preferentially be used with unfixed cells.

- After washing with sterile PBS, hepatocytes cultured or cytocentrifuged on coated-glass coverslips are incubated with Hoechst 33258 (5–10 μg/mL), for 10–30 min at 37°C in the dark.
- Wash hepatocytes with sterile PBS (*see* **Notes 2 and 3**). Nuclei can immediately be observed with the fluorescence microscope and images recorded for analysis.

A minimum of 500 nuclei have to be counted in several random fields to determine the percentage of apoptotic cells in the analyzed cell population. Positive controls for apoptosis should be used, as for instance hepatocytes treated with transforming growth factor β *(3)*.

The microscopic observation of apoptotic hepatocytes must reveal smaller size and highly fluorescent nuclei with condensed chromatin at the membrane level. Nucleolar dissolution can also be observed in some nuclei.

3.1.1.1.2. 4′-6-Diamino-2-phenylindole Dihydrochloride 4′-6-Diamino-2-phenylindole dihydrochloride (DAPI) has been documented to form fluorescent complexes with natural double-stranded DNA of intact and fixed cells. Like Hoechst dyes, DAPI is a blue fluorescent DNA stain that is considered to be stable especially for the DNA of fixed cells. Its absorption maximum is at 344 nm, whereas the emission maximum is at 449 nm (*see* **Note 4**).

- Hepatocytes grown or cytocentrifuged on glass coverslips are washed with sterile PBS and fixed with 4% paraformaldehyde for 20 min at room temperature.
- After washing three times with sterile PBS, hepatocytes are incubated with DAPI (0.2 μg/mL in PBS) or stored at 4°C for further analysis.
- DAPI incubation is performed for 10–25 min at room temperature in the dark.

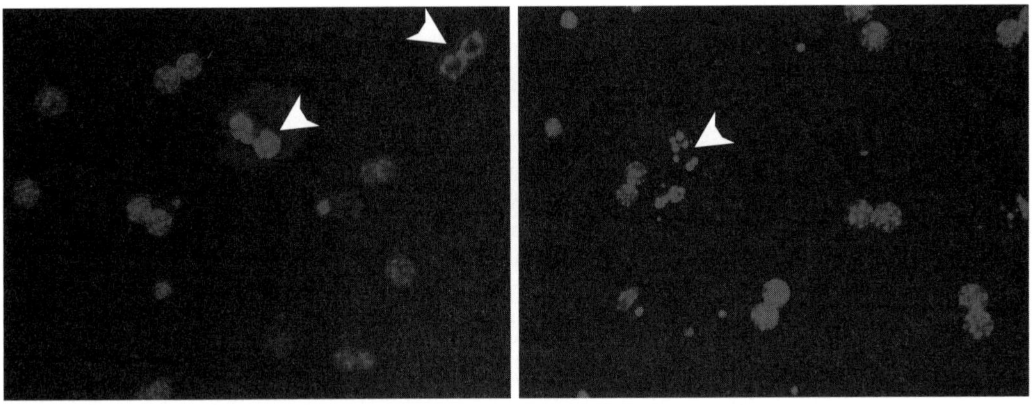

Fig. 6.1. Apoptotic nuclei and bodies observed in mouse primary hepatocyte cultures after staurosporine treatment (white arrows). Freshly isolated mouse hepatocytes were plated for 24 h on a collagen type I-coated coverslips in well plates and treated for 4 h with 1 μM staurosporine. Cells were thereafter fixed with 4% of formaldehyde for 20 min at room temperature, stained with DAPI for 30 min and analyzed using a fluorescence microscopy. (*see* Color Plate 1)

- Wash the hepatocytes three to five times with sterile PBS and coverslips are mounted on slides with Fluoprep (Biomérieux) or Mowiol and observed under a fluorescence microscope.

Appropriate controls must be used, for instance untreated and 4 h staurosporine (**Fig. 6.1**) (or other apoptosis-inducing agents) treated-primary hepatocytes.

3.1.1.1.3. Propidium Iodide PI is a cell-impermeant dye and is used to evaluate the proportion of dead cells within a cell population. After crossing the membranes of dead cells, PI binds the double-stranded DNA and red staining can be observed in the nucleus using fluorescence microscopy. Although this intravital dye stains damaged cells, it can also be used for the analysis of end-stage apoptotic cells. Its excitation complex at 535 nm with DNA absorption results in maximum emission at 516 nm. The association of this dye with Annexin V (*see* **Section 3.4**) is used to differentiate necrotic, apoptotic and living cells.

- Hepatocytes grown or cytocentrifuged on glass coverslips are washed with sterile PBS and fixed with 4% paraformaldehyde for 20 min at room temperature.
- Fixed hepatocytes are incubated with 10–20 μg/mL of PI solution for 10–30 min at room temperature.
- Wash cells with sterile PBS.
- Coverslips with hepatocytes are mounted on slides using Fluoprep or other antifade reagents and observed under a fluorescence microscope.

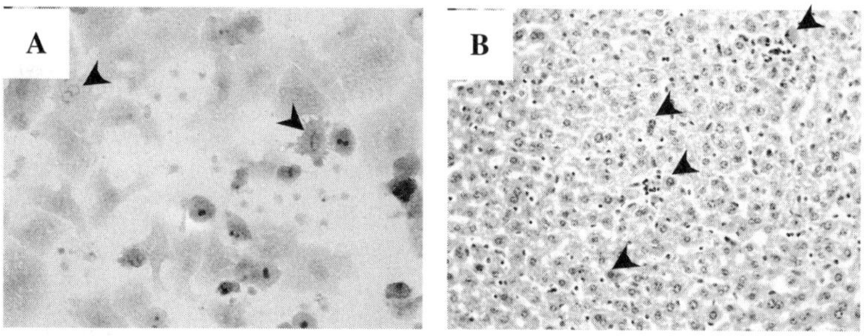

Fig. 6.2. Condensation of chromatin at the periphery of the nucleus in apoptotic mouse hepatocytes (black arrows). (A) Primary mouse hepatocytes were plated for 24 h in coated collagen type I well plates and treated for 4 h with 1 μM staurosporine. Cells were thereafter fixed with 4% formaldehyde for 20 min at room temperature and stained with HE for 10 min. (B) slice of mouse liver prefixed with formaldehyde, paraffin-embedded and HE-stained. (*see* Color Plate 2)

3.1.1.2. Non-fluorescent Markers

Hematoxylin–eosin (HE) staining is another approach to examine the presence of apoptotic cells. Hematoxylin and eosin stain the nucleus blue and the cytoplasm pink, respectively (*see* **Note 5**). For apoptotic cells, staining will reveal pycnotic nuclei with dense staining of chromatin, eosinophilic cytoplasm (see **Fig. 6.2**) and apoptotic bodies. HE staining is usually used for tissue sections analysis and, in contrast to fluorescent dyes, requires low-cost reagents, light microscope and microtome.

3.2. DNA Fragmentation

DNA fragmentation, a late-stage hallmark of the apoptotic process, is an irreversible biochemical event that occurs as a result of endonuclease-induced cleavage of nuclear DNA *(4)*. The obtained oligonucleosomal fragments with size of 180–200 base pairs can be visualized using different techniques. The analysis of DNA fragmentation is helpful when difficulties of DNA labeling are observed. The approach remains qualitative, as no precise information can be given regarding the amount of degraded DNA per cell. It is also moderately insensitive because of the low quality of recovered DNA. The approach may be assessed by radioactive and non-radioactive assays. In this chapter, we will focus only on non-radioactive assays. The technique is based on the lysis of hepatocytes to release the DNA, which, after precipitation and dissolution, can be directly loaded on agarose gels or spectrophotometrically analyzed using colorimetric assay.

3.2.1. Agarose Gel Electrophoresis

To recover DNA, several treatment protocols can be used. After extraction from hepatocyte lysate (*see* **Note 6**), the DNA can be loaded on 1–2% agarose gel and fragmentation is revealed by a ladder pattern due to DNA fragments. The timing of the assay can vary from 90 min to 2 days, whereas commercial kits without the extraction step (non-use of toxic reagents such as phenol, chloroform) are currently available. The technique is quite simple and requires low-cost equipment. However, it does not inform about the number of apoptotic cells within the analyzed cell population.

The standard protocol was described by Gong et al. *(5)* and did not use toxic reagents such as chloroform and phenol.

- Pellet-suspended hepatocytes after centrifugation at $200\times g$ for 10 min.
- Suspend the pellet in HBSS and incubate the cell suspension with phosphate-citrate buffer, pH 7.8 for 1 h at 37°C. This step allows extraction of low-molecular-weight DNA fragments after centrifugation.
- DNA is purified after cell lysis using 0.25% Nonidet P40 and treatment of the suspension with RNAse (1 mg/mL) and proteinase K (2 mg/mL).
- Fifteen micrograms of extracted DNA is loaded on 1–2% agarose gel and electrophoresed at 100 V for 2 h. Detection of the DNA is performed using EB and UV light (*see* **Note** 7).

Other original experimental procedures have been documented *(6)* and adapted to hepatocyte suspension *(7)* but used solvents such as phenol and chloroform. Many steps of the procedure are critical for the analysis of DNA fragmentation, leading to adapt several parameters for instance extraction and purification steps and time of DNA precipitation and dissolution (*see* **Notes 8–10**).

3.3. TUNEL Assay

TUNEL technique was developed to track the apoptotic cells in situ *(8)*. The technique is based on the transfer of nucleotides, catalyzed by TdT, on the free 3× OH ends of the cleaved DNA. The insertion of tagged-nucleotides can be revealed by specific antibodies. The technique is sensitive, more specific for apoptosis, can be assessed simultaneously with the analysis of morphology and allows the detection of a small number of apoptotic cells within the examined cell population. For reproducibility, accuracy and reducing time, it is highly recommended to use commercially available kits with appropriate controls. Regarding liver cells, it has been documented to be aware of the false positive that could be obtained as demonstrated in mouse hepatocytes and rat liver tissues *(9, 10)*.

For isolated hepatocytes:

- Hepatocyte smears, adherent on coated substrate according to the revelation system used (fluorescence or colorimetry) or cytocentrifuged hepatocytes, are fixed using 4% paraformaldehyde for 20 min at room temperature (*see* **Note 11**).
- After washing with sterile PBS, hepatocytes are treated with proteinase K (15 min at 37°C), permeabilized with 0.1 %

Triton X-100 (2 min on ice) and incubated with TdT and conjugated nucleotides.

- Inserted nucleotides are revealed using specific antibodies and microscopes (*see* **Note 12**).
- Nuclei are counterstained with HE and DAPI for non-fluorescent and fluorescent techniques, respectively.
- Save the digital images of several fields and score the apoptotic cells within the analyzed cell suspension (*see* **Note 13**).

For tissue sections:

- If frozen, slices are fixed whereas the paraffin-embedded ones are dewaxed and rehydrated as in standard protocols (*see* **Note 14**).
- Slices are treated for deproteinization (proteinase K, 20 μg/mL 15 min at 37°C) and permeabilization before incubation with TdT and coupled nucleotides.
- Inserted nucleotides are revealed using specific antibodies and microscopes.

3.4. Flow Cytometry

The procedure allows the quantification of fluorescence intensity per cell within a cell population. Such a sensitive approach remains complementary to microscopic evaluation and may supply more rapid and accurate data than manual counting. However, it needs high-cost equipment and is not adapted for tissue or tissue-cultured cells. Because apoptosis could alter all the cell compartments according to the cell death stages, strategies based on flow cytometry were developed to analyze this phenomenon at the membrane, cytoplasmic, mitochondrial and nuclear levels. As early apoptosis may be related to membrane permeability changes, fluorescent dyes (*see* **Section 3.1.1.1**) could also be used in flow cytometry, allowing the rapid quantification of apoptotic permeant cells in a large population.

According to hepatocyte size variability, light scattering cannot be usefully used for apoptosis evaluation. Annexin V staining, which informs about the phospholipid-like phoshatidylserine asymmetry (which could be lost before membrane integrity) in the cell membrane, allows the analysis of early apoptotic cells within the cell population. Its combination with nuclear dyes (as PI), which helps for the discrimination of early and late apoptotic cells, was used for hepatocyte analysis *(11, 12)*. Another advantage is the short time required for cell staining and data analysis.

- Hepatocytes ($2.5–5 \times 10^5$) are suspended in specific buffer (HBSS, PBS) or medium and are centrifuged at $1000 \times g$, 5 min at room temperature.
- Discard the supernatant and suspend the pellet in 0.5 mL of cold sterile PBS.

- Centrifuge the cells at $1000\times g$ 5 min at room temperature.
- Discard the supernatant and suspend the pellet in 95 μL of Annexin binding buffer 1× (*see* **Note 15**).
- Add 5 μL of labeled Annexin as recommended by BD Bioscience.
- Incubate for 15 min at room temperature in the dark.
- Analyze the cell suspension in the flow cytometer.

3.5. Determination of Mitochondrial Membrane Potential

Mitochondria play a critical role in the regulation of apoptotic cell death by mechanisms that are conserved through evolution. Mitochondria maintain ATP production *(12)*, mitochondrial membrane potential ($\Delta\psi$) and permeability *(13, 14)*.

$\Delta\psi$, the electrochemical gradient across the mitochondrial membrane, is an indicator of mitochondrial activity and membrane integrity. Its depolarization induces the release of apoptotic proteins to the cytosol *(15)*. It can be analyzed using membrane lipophilic cationic probes, which can be accumulated inside the mitochondria because of the negative inside membrane potential. Rhodamine 123 was widely used for hepatocytes and can be used in association with Annexin to clearly distinguish apoptotic dead and living cells within the cell population *(16)*. The technique is quite simple and data can be analyzed using a fluorimeter or a flow cytometer.

How to proceed:

- Incubate 2×10^6 hepatocytes with 1 μM Rhodamine 123 suspended in Williams' or other medium for 10 min at 37°C in the dark and with agitation (*see* **Note 16**).
- Wash hepatocytes three times with a double volume of sterile PBS before monitoring the fluorescence (excitation and emission wavelengths of 498 and 524 nm, respectively) (*see* **Note 17**).

3.6. Analysis of Cytoplasmic Cell Compartment

Another feature of hepatocyte apoptosis study is the analysis of cytoplasmic compartment because intracellular apoptotic pathways have been well characterized and described. The analysis of this cell compartment is important because cytoplasmic apoptotic pathways are independent from those acting in the nucleus. Proteins extracted from hepatocyte cytoplasm can be directly analyzed at the levels of expression and activity. The study of the subcellular expression of specific markers can also inform about the apoptosis induction but the data remain correlative and have to be confirmed by complementary analyses.

3.6.1. Release of Cytochrome C

Cytochrome C remains one of the well-studied markers besides caspases. The cytosolic release of this electron-transporting protein of the mitochondrial peripheral membrane leads to the activation of caspase 3 and the induction of apoptosis. The expression of cytochrome C could be analyzed by specific antibodies using Western blotting after cellular fractionation and immunocytochemistry.

The first step of the procedure consists in homogenizing hepatocytes using chemicals, enzymes or sound waves (sonication process). The obtained break-open hepatocyte suspension is submitted to centrifugation (in some cases silicon can be used) for the separation of mitochondria from the rest of the cytoplasm.

- Pellet the cell suspension containing 1.5×10^7 hepatocytes after low-speed centrifugation and wash the cells with PBS.
- Hepatocytes are suspended and incubated for 2 min at room temperature in 0.8 mL of permeabilization medium.
- In a polypropylene microtube, 500 μL of the homogenized hepatocyte suspension are layered on the top of a silicon oil layer (800 μL).
- Centrifuge the permeabilized hepatocytes through the silicon oil layer for 30 s at $13,500 \times g$ into 250 μL of 250 mM sucrose solution.
- Recover the upper part that contains the cytosolic fraction and freeze it at –80°C until analysis.
- Suspend the mitochondria pellet of the lower fraction with 200 μL of lysis buffer and incubate for 10 min on ice.
- Centrifuge the mitochondria lysate for 2 min at $13,500 \times g$ at 4°C.
- Recover 180 μL of the supernatant and store at –80°C for further analysis.
- Dose the protein concentration and analyze with SDS-PAGE.
- Fifty micrograms of the total extracted proteins from each compartment were separated by SDS-PAGE and transferred to nitrocellulose membranes.
- Use a specific primary antibody for western blotting (*see* **Notes 18 and 19**).

3.6.2. Caspase Activity

Caspases are aspartate-specific cysteine proteases that use the sulfur atom in cysteine to cleave polypeptide chains. They initially exist as pro-caspases and are activated as a consequence of the propagation of a death-inducing signal, leading to the cleavage of several intracellular substrates ensuring cell death. For instance, nuclear shrinkage occurs due to the caspase cleavage of nuclear lamins, cellular shape loss is due to the cleavage of cytoskeletal proteins whereas loss of cell adherence is the result of caspase attack on focal adhesion kinase *(17)*. Regarding liver cell transplantation, the involvement of caspase-3 activity, an initiator of apoptotic early events, has already been demonstrated for the induction of hepatocyte apoptosis both in vitro and in vivo *(18, 19)*.

Caspase involvement can be evaluated at the level of its expression or activity using immunological techniques, such as western blotting, flow cytometry, immunocytochemistry and immunohistochemistry. The advantage of the latter is the possible combined analysis of cell morphology and apoptotic markers expression and/or activity. Biochemical assays are quite simple and may quantitatively evidence the binding of labeled peptide on the active site of the caspase or the formation of labeled products after the cleavage of substrate. It is also possible to detect substrate cleavage using western blotting.

Measurement of caspase activity by evaluating the proteolytic cleavage of specific labeled substrates is very useful for the detection and quantification of apoptosis. Fluorescence-, absorbance- or luminescence-based assay kits are available and ready to use for the rapid analysis of caspase activity. This can be evaluated both on hepatocyte lysates and in situ, leading to select the appropriate assay for apoptosis detection sensitivity (*see* **Note 20**). The data can be presented as percentage or fold increase vs control samples. The protein lysis step is crucial to avoid contamination by other proteases. Appropriate positive and negative controls should accordingly be used to evaluate both the efficacy and the specificity of the assay. These assays are only suitable for hepatocyte suspensions or primary culture. After liver cell transplantation, evaluation of caspase activity should be analyzed in situ to corroborate the detection of engrafted donor cells and apoptosis within the recipient liver parenchyma.

- Suspended hepatocytes are centrifuged at $666\times g$, 3 min at 4°C.
- Remove the supernatant and re-suspend the pellets in cell lysis buffer.
- Hepatocyte lysates are transferred to 96-well flat-bottomed plates, incubated for 30 min at 4°C (it is recommended to microscopically confirm the cell lysis) and re-centrifuged at $3838\times g$ for 5 min (to eliminate nuclei).
- Fifty micrograms of the total extracted proteins were incubated with the labeled substrate for 1 h and the resulted fluorescence or absorbance was measured.

3.7. Transmission Electron Microscopy

In this paragraph, we will only focus on the analysis of isolated hepatocyte suspension. Transmission electron microscopy (TEM) is a very slow procedure and, in contrast to light microscopy, needs more hepatocytes. It also needs very experienced people for accurate analysis of the data. Therefore, the technique cannot accordingly be used as a routine assay for apoptosis evaluation on hepatocytes. Because of the small number of cells analyzed per section, TEM can neither be used for quantitative evaluation of hepatocyte apoptosis. However, TEM remains more adapted for tissue analysis and can supply appropriate controls and complementary data to the other apoptotic evaluation tests cited above.

- Samples of suspended hepatocytes are centrifuged at 1200 r.p.m. for 5 min at room temperature.
- After supernatant removal, hepatocyte pellets were fixed with 2.5% EM grade glyceraldehyde buffered in 0.1 M sodium cacodylate, for 48 h at 4°C and post-fixed in 1% osmium tetroxide (*see* **Notes 21 and 22**).
- After embedding in Epoxy Embedding Medium, semi-thin sections were contrasted with uranyl acetate and lead citrate (*see* **Note 23**) before examination using TEM at a magnification of ×4140.

4. Notes

1. The decision as to which technique to use for the evaluation of apoptosis in the context of hepatocyte transplantation is based both on the specific addressed question and on the time schedule, especially for hepatocyte suspension (after isolation and thawing) dedicated to immediate transplantation.
2. No permeabilization step is needed for Hoechst staining. The compound is dissolved in H_2O (precipitation with PBS) at 1 mM concentration. Working solution should be prepared fresh prior to each assay by diluting the dye in warmed buffer.
3. Incubation time is determined depending on the transport efficiency of the dye and staining kinetics should be adapted to experimental conditions. pH and NaCl concentrations are also determinant for the binding of Hoechst to the DNA.
4. DAPI staining of living cells is slow to appear whereas in fixed cells the dye can, in certain conditions, form complexes with other cellular compounds as RNA and tubulin.
5. For more details, see http://www.ihcworld.com/_protocols/special_stains/h&e_ellis.htm
6. Lysis time must be determined depending on the sample used (liver tissue or hepatocyte suspension).
7. Loading buffer is used to easily load the wells of the agarose gel and to follow the samples electrophoresis (bromophenol blue dye) whereas EB (carcinogenic agent!) will stain DNA

for visualization on a UV transilluminator (eye and skin protection).

8. Hepatocytes, transcriptionally active cells, contain high levels of RNA, which could be co-extracted with DNA. Hepatocyte lysate should be treated with RNAse to digest the contaminating RNA.
9. In some primary cell cultures, spontaneous cell DNA fragmentation can occur, leading to an increased background.
10. The dissolution of extracted DNA, which depends on the recovered quantity and the size of the analyzed sample, is crucial as non-fragmented DNA may need higher volumes of TE buffer.
11. This step is important to avoid the loss of low-molecular DNA fragments during the permeabilization step.
12. For non-fluorescent detection, inactivation of endogenous enzymes is recommended for a lower background. Such information could also be obtained after incubation of the sections without TdT.
13. In some experimental conditions, apoptosis is not accompanied by DNA degradation and vice versa, leading to use of the TUNEL assay in parallel to other techniques such as morphological analyses.
14. Slices are placed in successive solutions of xylene, methanol and tap water. From that step, never let slices dry out.
15. Labeled Annexin binding is sensitive to salts and calcium concentration.
16. Rhodamine 123 (red powder) is poorly dissolved in H_2O. Dilution can be performed in ethanol at 1 mg/mL (stored at –20°C for several months in the dark), whereas H_2O can be thereafter used for intermediate dilutions. Shaking will facilitate the incorporation of the probe, which higher concentrations (>1 mM) may inhibit $F0$–$F1$ ATPase and mitochondrial respiration.
17. For labeling specificity, the analysis of $\Delta\psi$ in the presence of mitochondrial depolarizing agents such as dinitrophenol is highly recommended.
18. For cytochrome C detection using western blotting, appropriate controls should be used (apoptotic and non-apoptotic cell extracts). To specifically evaluate the compartmentalization of cytochrome C, expression of extra-mitochondrial proteins such as actin should be analyzed and serves as the control of the purity of the mitochondrial and extra-mitochondrial protein fractions.
19. For immunocytochemistry, specific mitochondrial dyes should be used to accurately evaluate the intracellular

expression pattern of cytochrome C. In non-apoptotic cells, mitochondrial staining of cytochrome C should reveal a punctuate signal that coincides with the dye staining. In apoptotic cells, cytoplasmic release of cytochrome C coincides with its instability, leading them to lose staining in some cases.

20. Titration should be studied to evaluate the limit of detection of the assays used especially for cell lysates. Intermediate dilutions of initial protein extracted from hepatocytes suspension should be carefully performed and it depends on the total volume of the biochemical reaction.
21. Fixation step of the samples is crucial for ultrastructure preservation and must be performed in a fume hood.
22. Osmium tetroxide needs at least 24 h to completely dissolve. Hence, the stock solution (4%) must be prepared in advance. According to the thickness of the sections, the percentage of osmium tetroxide used must be adapted. One percent is usually used for cell pellets.
23. Lead citrate commonly used for section counterstaining can be prepared as follows: add 4.8 mL of double-distilled water to 0.133 g of lead nitrate and shake gently to dissolve. Add 0.176 g of trisodium citrate until a milky solution is formed before adding 200 μL of 4 M NaOH.

References

1. Zvibel, I., Smets, F., Soriano, H. (2002) Anoikis: roadblock to cell transplantation? *Cell Transpl* 11, 621–630.
2. Tanaka, K., Soto-Gutierrez, A., Navarro-Alvarez, N., et al. (2006) Functional hepatocyte culture and its application to cell therapies. *Cell Transpl* 15, 855–864.
3. Gressner, A. M., Lahme, B., Mannherz, H. G., et al. (1997) TGF-beta-mediated hepatocellular apoptosis by rat and human hepatoma cells and primary rat hepatocytes. *J Hepatol* 26, 1079–1092.
4. Wyllie, A. H., Kerr, J. F., Currie, A. R. (1980) Cell death: the significance of apoptosis. *Int Rev Cytol* 68, 251–306.
5. Gong, J., Traganos, F., Darzynkiewicz, Z. (1994) A selective procedure for DNA extraction from apoptotic cells applicable for gel electrophoresis and flow cytometry. *Anal Biochem* 218, 314–319.
6. Sambrook, J., Fritsch, E. F., Maniatis, T. (1989) *Molecular Cloning: A laboratory Manual,* 2nd edn. Cold Spring Harbor Laboratory Press, Cold Spring Harbor, NY.
7. Smets, F. N., Chen, Y., Wang, L. J., et al. (2002) Loss of cell anchorage triggers apoptosis (anoikis) in primary mouse hepatocytes. *Mol Genet Metab* 75, 344–352.
8. Gavrieli, Y., Sherman, Y., Ben Sasson, S. A. (1992) Identification of programmed cell death in situ via specific labeling of nuclear DNA fragmentation. *J Cell Biol* 119, 493–501.
9. Pulkkanen, K. J., Laukkanen, M. O., Naarala, J., et al. (2000) False-positive apoptosis signal in mouse kidney and liver detected with TUNEL assay. *Apoptosis* 5, 329–333.
10. Stahelin, B. J., Marti, U., Solioz, M., et al. (1998) False positive staining in the TUNEL assay to detect apoptosis in liver and intestine is caused by endogenous nucleases and inhibited by diethyl pyrocarbonate. *Mol Pathol* 51, 204–208.
11. Fu, T., Blei, A. T., Takamura, N., et al. (2004) Hypothermia inhibits Fas-mediated

apoptosis of primary mouse hepatocytes in culture. *Cell Transplant* 13, 667–676.
12. Stephenne, X., Najimi, M., Khuu, N. D., et al. (2007) Cryopreservation of Human Hepatocytes Alters the Mitochondrial Respiratory Chain Complex 1. *Cell Transpl* 16, 409–419.
13. Vayssiere, J. L., Petit, P. X., Risler, Y., et al. (1994) Commitment to apoptosis is associated with changes in mitochondrial biogenesis and activity in cell lines conditionally immortalized with simian virus 40. *Proc Natl Acad Sci USA* 91, 11752–11756.
14. Kroemer, G., Reed, J. C. (2000) Mitochondrial control of cell death. *Nat Med* 6, 513–519.
15. Mignotte, B., Vayssiere, J. L. (1998) Mitochondria and apoptosis. *Eur J Biochem* 252, 1–15.
16. Ly, J. D., Grubb, D. R., Lawen, A. (2003) The mitochondrial membrane potential (deltapsi(m)) in apoptosis; an update. *Apoptosis* 8, 115–28.
17. Rudel, T., Bokoch, G. M. (1997) Membrane and morphological changes in apoptotic cells regulated by caspase-mediated activation of PAK2. *Science* 276, 1571–1574.
18. Yagi, T., Hardin, J. A., Valenzuela, Y. M., et al. (2001) Caspase inhibition reduces apoptotic death of cryopreserved porcine hepatocytes. *Hepatology* 33, 1432–40.
19. Song, E., Chen, J., Antus, B., et al. (2001) Adenovirus-mediated Bcl-2 gene transfer inhibits apoptosis and promotes survival of allogeneic transplanted hepatocytes. *Surgery* 130, 502–11.

Chapter 7

Small Animal Models of Hepatocyte Transplantation

Jurgen Seppen, Ebtisam El Filali, and Ronald Oude Elferink

Abstract

In this chapter, we describe techniques used to determine the efficiency of hepatocyte transplantation in animal models of liver disease. We have included the Gunn rat as a model of an inherited liver disease without hepatocyte damage and *Abcb4* knockout mice as a model for an inherited liver disease with hepatocyte damage. Immunodeficient mice are included as an animal model for human hepatocyte transplantation.

We describe problems that can be encountered in the maintenance and breeding of Gunn rats and immunodeficient *Rag2/gamma common* knockout mice. Protocols for the collection of bile in rats and mice are described, and we have also detailed the detection of green fluorescent protein (GFP)-labelled human hepatocytes in immunodeficient mice in this chapter.

Keywords: Gunn rat, bilirubin, bile collection, *UGT1A1*, *Abcb4*, *PFIC3*, Crigler–Najjar, glucuronyltransferase, liver.

1. Introduction

The first studies on liver transplantation in animal models showed that this procedure was feasible but also revealed that considerable morbidity and mortality occurred *(1)*. The transplantation of hepatocytes instead of whole livers was therefore already considered in an early stage. The first experimental model used in the development of liver cell transplantation was the Gunn rat. This strain of rats is the model of Crigler–Najjar disease and is characterised by the absence of the hepatic enzyme bilirubin *UDP glucuronyltransferase*. Because Gunn rats are not able to conjugate bilirubin with glucuronic acid, high concentrations of toxic bilirubin occur in the circulation.

Transplantation of normal hepatocytes into the portal vein of Gunn rats was shown to partially correct the hyperbilirubinaemia

Anil Dhawan, Robin D. Hughes (eds.), *Hepatocyte Transplantation, vol. 481*

DOI 10.1007/978-1-59745-201-4_7 Springerprotocols.com

for up to 12 weeks *(2)*. The Gunn rat model has subsequently been used in several studies designed to optimise hepatocyte transplantation procedures, culminating in the treatment of Crigler–Najjar patients by this procedure *(3)*. What has become clear from these studies is that the grafting efficiency of hepatocyte transplantation is low. Whereas this low grafting efficiency may be sufficient in the treatment of inherited liver diseases that require minimal expression of the defective gene, other disorders would require a much larger liver cell replacement.

The liver has a remarkable regenerative capacity; after removal of up to 70% of the liver, normal liver mass is restored within 2 weeks. When the liver is damaged by a genetic deficiency or toxic substance, transplanted hepatocytes that are resistant to this damage will have a growth advantage and can preferentially repopulate the liver. This phenomenon has been first described in the urokinase plasminogen activator transgenic mouse. These mice exhibit severe liver damage; transplantation of these mice with normal hepatocytes leads to virtually complete repopulation of the liver with the donor cells *(4)*. Several disease models exist in which hepatocytes are damaged by a genetic deficiency. Fumarylacetoacetate hydrolase *(Fah)* deficiency causes accumulation of fumarylacetoacetate and/or maleylacetoacetate, which results in severe liver damage. After transplantation of *Fah*-deficient mice with normal liver cells, repopulation of the host liver with transplanted cells will take place *(5)*.

Another model in which liver cell repopulation can occur is the deficiency of the canalicular phosphatidylcholine (PC) transporter *Abcb4*. The excretion of PC serves to inactivate the detergent activity of high concentrations of bile salts present in bile. The absence of *Abcb4* causes progressive familial cholestasis type 3. Mice with *Abcb4* deficiency suffer from mild progressive liver disease; feeding these animals a diet containing the bile salt cholic acid strongly aggravates liver damage. Because the deficiency of *Abcb4* causes hepatocyte toxicity, normal liver cells have a growth advantage in *Abcb4* knockout mice. Transplantation of normal hepatocytes into *Abcb4* knockout mice leads to partial repopulation of the liver by these cells *(6)*.

These animal models of liver cell repopulation are clinically relevant since a recent paper shows that repopulation of the liver with transplanted normal cells will also occur in humans suffering from a genetic deficiency that damages the liver cells *(7)*. It is therefore also important to have an animal model in which transplantation of human hepatocytes can be studied. One of the best immune-deficient models are mice with disrupted *Rag2* and interleukin receptor gamma common chain genes. The consequence is of this double knockout is a total absence of T, B and NK cells. These mice are better hosts for human tissues than

Scid or *Rag1/2* knockout mice, which may have some NK cell activity. Another advantage of this strain is that they do not spontaneously develop tumors, which makes long-term studies possible.

2. Materials

2.1. Collection of Bile from Gunn Rats

1. 1 ml syringe and 25 gauge needles 5/8 (0.5×16 mm).
2. Operation instruments: Scissors (Medicon 03.06.14, 02.04.10), dissecting forceps, tissue forceps, vessel clip, self-retaining retractors, hooked sharp forceps (Aesculap, BD 501, BD 216, FE 13 K, BV74, BD329). Microscissors (Moria 9600). Hook (Aesculap Brom BT75).
3. Canule (Venencatheter, 0.5×0.9 mm, B. Braun).
4. Suture material (Ethicon 5-0, EH781).
5. Eppendorf vessels, sterile gauze, cotton tips and blood absorption swabs.
6. Anaestetic, Nembutal (sodiumpentobarbital, 60 mg/ml, Sanofi).

2.2. Collection of Bile from Abcb4 *Knockout Mice*

1. 1 ml syringe and 25 gauge needles 5/8 (0.5 × 16 mm).
2. Operation instruments: scissors, tissue forceps (Medicon 02.10.10, 06.30.10), dissecting forceps, self-retaining retractors, hooked-sharp forceps, hooked forceps (Aesculap, BD 501, BV74, BD329, OC22); Microscissors (Moria 9600); Hook (Aesculap Brom BT75).
3. Canule (polyethylene, 0.4 × 0.8 mm, Portex Limited).
4. Suture material (Ethicon 5-0, EH781).
5. Eppendorf vessels, sterile gauze, cotton tips and blood absorption swabs.
6. FFD mix for anaesthesia: 4.5 ml 0.9% NaCl + 0.3 ml Hypnorm (10 mg/ml fluanisone, 0.315 mg/ml fentanyl citrate) + 0.3 ml diazepam (5 mg/ml) (Janssen Pharmaceutica, Beerse, Belgium).

2.3. Fixation of Intact Animals for Direct Fluorescence Detection of Transplanted GFP-Positive Cells

1. Phosphate-buffered saline (PBS), 30% sucrose solution.
2. Paraformaldehyde (PFA) in PBS, 2 and 4%. The solutions needs to be heated to 70°C in order to dissolve the PFA and must then be cooled to room temperature before use. The solution may be stored at –20°C.
3. Infusion set (Microflex: 0.5 mm, 25G, Vygon 246.05).
4. Scissors, dissecting forceps (Medicon 02.10.10, 06.30.10).

5. Freezing vials.
6. FFD mix for anaesthesia: 4.5 ml 0.9% NaCl + 0.3 ml Hypnorm (10 mg/ml fluanisone, 0.315 mg/ml fentanyl citrate) + 0.3 ml diazepam (5 mg/ml) (Janssen Pharmaceutica Beerse, Belgium).

2.4. Preparation of Cryosections of Fixed Livers on Poly-L-Lysine-Coated Glass Slides

1. Poly-L-lysine stock solution 10 mg/ml poly-L-lysine (Sigma, P-1399) in bidistilled water). Store aliquots of the stock at –20°C. Dilute the stock solution of poly-L-lysine prior to use to a final concentration of 0.1 mg/ml (1:100) using 10 mM Tris-HCL (pH 8.0).
2. Microtome suitable for cryosectioning.
3. Embedding medium: tissue-tek OCT compound (Bayer 4583).
4. Disposable microtome blades (model S35, Klinipath, 02.075.00.000).
5. Mounting medium (Vectashield, Vectorlabs H-1200).

3. Methods

Because transplantation and histochemical techniques are already covered in other chapters of this volume, we will describe techniques used to determine transplantation efficiency in Gunn rats (*see* **Note 1**), *Abcb4* knockout mice and immune-deficient mice (*see* **Note 2**).

In animal models of inherited liver diseases in which biliary excretion of compounds is affected, it is important to collect bile to determine the therapeutic efficiency of hepatocyte transplantation. We therefore describe techniques to collect bile from mice and rats.

Detection of human cells in murine liver can be difficult. One of the easiest ways is to mark the human cells with green fluorescent protein (GFP). This can be done by transduction with GFP lentiviral vectors as described elsewhere in this volume. We therefore include a protocol for the detection of GFP-labelled liver cells by direct fluorescence microscopy.

3.1. Collection of Bile from Gunn Rats

1. Weigh the rat and give the anaesthetic (0.1 ml nembutal per100 g bodyweight, intraperitoneally).
2. Shave the belly and open the skin and the peritoneal cavity.
3. Spread the wound, take the intestine out of the peritoneal cavity and position it to the left side of the rat. Cover the external intestine with sterile gauze wetted with saline.

4. Cut the membrane between the liver and the diaphragm and position the liver.
5. Put an atraumatic vessel clip on the duodenum, the bile duct will be visible as a thin white line.
6. Carefully put a ligature (ethicon 5-0) around the bile duct at the caudal side and tie it (do not cut away the loose ends).
7. Clean the bile duct carefully from unwanted tissues (pancreatic tissue, fatty tissue).
8. Put a ligature (ethicon 5-0) around the bile duct at the cranial side. Make one knot but do not tie it yet.
9. Make a cut, using the microscissors in the bile duct between the two ligatures and keep it open with a hook.
10. Put the cannula in the bile duct, push it towards the liver but keep it distal from the bifurcation.
11. First tie the cranial suture, then tie the caudal suture.
12. Position the cannula and put the end into a collection vessel.
13. Protect the cannula and collection vessel from light by covering it with aluminum foil (*see* **Note 3**).

3.2. Collection of Bile from Abcb4 *Knockout Mice Fed a Cholate Diet*

1. Administer the FFD anaesthetic to the mouse (100 μl FFD mix per 5 g bodyweight, intraperitoneally).
2. Shave the belly and open the skin and the peritoneal cavity. Spread the wound, take the intestine out of the peritoneal cavity and position it to the left side of the mouse. Cover the external intestine with sterile gauze wetted with saline.
3. Cut the membrane between the liver and the diaphragm and position the liver.
4. Ask someone to lift the xyphoid to enhance visibility of the gallbladder.
5. Put a ligature around the bile duct between the gallbladder and the duodenum and tie it.
6. Put a ligature around the gallbladder with one double knot but do not tie it yet. To get a better view, use magnifying glasses.
7. Pick up the gallbladder at the tip and cut a small hole at the top of the bladder using the microscissors.
8. Insert the cannula and tie the ligature with the double knot, then tie two single knots.
9. Position the cannula for optimal flow and put the intestine back in the abdomen.
10. Cut the cannula for optimal contact with the collection vessel and to create a better flow.
11. Add 100 μl of FFD mix on top of the intestine to maintain the right level of anaesthesia.

3.3. Fixation of Intact Animals for Direct Fluorescence Detection of Transplanted GFP-Positive cells

1. Administer the FFD anaesthetic to the mouse (100 μl FFD mix per 5 g bodyweight intraperitoneally).
2. Make an incision over the entire abdomen using the surgical scissors.
3. Make sure all equipment is laid out next to you as the following steps will require to be performed as quickly and smoothly as possible.
4. Carefully cut the thorax open along the sternum. Make sure the thorax is flapped to the sides so that the heart can be well viewed.
5. Insert the needle of the infusion set in the apex of the heart.
6. Cut the vena cava inferior, proximally situated from the liver to ensure good perfusion.
7. Perform an intracardial perfusion with 20 ml PBS in approximately 1 min. The liver should become pale soon after the start of the perfusion.
8. Change the syringe to one containing 20 ml 2% PFA. Upon 2% PFA, perfusion the body of the mouse will become rigid.
9. The perfused tissues of interest are dissected out and further fixed for 2–4 h in 4% PFA at room temperature.
10. Fixed organs are incubated overnight in 30% sucrose at 4°C.
11. Cut the organs in smaller pieces prior to snap freezing them to facilitate the sectioning.
12. Place the tissues in cryotubes, snap-freeze them in liquid nitrogen and store at –80°C.

3.4. Preparation of Cryosections of Fixed Livers on Poly-l-Lysine-Coated Glass Slides

1. Soak glass slides overnight in 1% NaOH.
2. Rinse them the next morning for 15 min in running warm tap water followed by rinsing them briefly with distilled water.
3. Soak the slides for at least 1 h in 2% HCL and rinse again for 15 min in running warm tap water and briefly with Elix water.
4. Place the glass slides in racks in a solution of 0.1 mg/ml poly-L-lysine. Incubate for 30 min at room temperature.
5. Dry the slides first in an air flow for 2–3 h followed by overnight placement in an incubator at 37°C.
6. The slides can be stored at room temperature.
7. Take the vials containing the liver samples to the cryostat on dry ice or in liquid nitrogen.
8. Make sure the working temperature of the cryostat is –24°C.
9. Apply sufficient amount of embedding medium on the specimen disc, avoid air bubbles and let it cool without solidifying. Place the frozen tissue sample on the embedding medium and let it equilibrate for at least 5 min.

10. Make sections with a thickness of 5 μm.
11. After sectioning, immediately attach the section on the poly-L-lysine-coated glass slide, which must be at room temperature.
12. Dry the sections briefly and add a drop of vectashield mounting medium containing DAPI on the sections and cover them with glass coverslips. The sections are now ready to be viewed under the fluorescence microscope (*see* **Note 4**).

4. Notes

1. Breeding and maintenance of Gunn rats. In some centers Gunn rats are bred as heterozygotes due to the severe phenotype of homozygous animals. However, we are able to breed homozygous mutant rats. A crucial factor in the breeding of Gunn rats is the chow used, we routinely fed the rats Hope Farms SRM-A chow. On this diet, bilirubin levels are generally below 150 μM. On some diets, serum bilirubin will be considerable higher; switching the rats to a purified diet (normal purified diet, Hope Farms) or to the Harlan Teklad 2018 diet caused a twofold increase in serum bilirubin. Breeding of rats fed Harlan Teklad 2018 diet was difficult because the newborn rats were killed by the mothers or had to be terminated because they appeared to have neurological damage. In contrast, Gunn rats maintained on SDS CRM(E) diet did not have an increased serum bilirubin as compared to Hope Farms SRM-A. However, Gunn rats on SDS CRM(E) diet did not reproduce. These observations indicate that the choice of diet is very important in the maintenance of Gunn rats and changes in diet should be tried if problems in maintenance or breeding of Gunn rats occur. Because Gunn rats are deficient in detoxification, they can be more sensitive to drugs commonly used in other rodents. For surgical procedures and drawing of blood isoflurane gas, anaesthesia is therefore preferred. For end-point procedures intraperitoneal injection of sodium pentobarbital can be used.
2. Breeding and maintenance of *RAG gamma common* knockout mice. Because these mice are immunodeficient, they are vulnerable to infections. Breeding of the mice is therefore preferably done in isolator devices or in individually ventilated cages. However, for experiments with an end point within half a year, the mice can be maintained in normal cages with filtertops.
3. Collection of bile. Bilirubin is very light sensitive, collection of Gunn rat bile to determine output of bilirubin should

therefore be performed with the canula and collection vessel covered with aluminium foil.

For canulations in mice and rats: try to make sure the sharp ends of the cannula are removed by rolling it in your fingers. Otherwise the sharp end may rupture the bile duct.

4. Preparation of cryosections of fixed livers on Poly-L-lysine-coated glass slides. The fluorescence of GFP is rapidly lost when unfixed livers are cryosectioned. Embedding of fixed tissue according to standard histochemical techniques in media such as paraplast also leads to loss of GFP fluorescence. Because cryosectioning of formaldehyde fixed livers is very difficult it is necessary to saturate the tissue samples with a 30% sucrose solution to facilitate sectioning. Because sucrose saturation makes liver sections prone to detachment from the glass slides it is subsequently necessary to use poly-L-lysine-coated slides to allow better attachment. Autofluorescence can be a problem in detecting GFP fluorescence in liver. If possible use a microscope equipped with a broad band emission filter for the detection of green fluorescence. GFP will fluoresce bright green whereas the autofluorescence will show up as yellow.

References

1. Starzl, T. E., Marchioro, T. L., Faris, T. D. (1966) Liver transplantation. *Ann Intern Med* 64(2),:73–477.
2. Matas, A. J., Sutherland, D. E., Steffes, M. W., et al. (1976) Hepatocellular transplantation for metabolic deficiencies: decrease of plasms bilirubin in Gunn rats. *Science* 192(4242), 892–894.
3. Fox, I. J., Chowdhury, J. R., Kaufman, S. S., et al. (1998) Treatment of the Crigler–Najjar syndrome type I with hepatocyte transplantation. *N Engl J Med* 338(20), 1422–1426.
4. Rhim, J. A., Sandgren, E. P., Degen, J. L., et al. (1994) Replacement of diseased mouse liver by hepatic cell transplantation. *Science* 263(5150), 1149–1152.
5. Overturf, K., Al Dhalimy, M., Tanguay, R., et al. (1996) Hepatocytes corrected by gene therapy are selected in vivo in a murine model of hereditary tyrosinaemia type I. *Nat Genet* 12(3), 266–273.
6. De Vree, J. M., Ottenhoff, R., Bosma, P. J., et al. (2000) Correction of liver disease by hepatocyte transplantation in a mouse model of progressive familial intrahepatic cholestasis. *Gastroenterology* 119(6), 1720–1730.
7. Stephenne, X., Najimi, M., Sibille, C., et al. (2006) Sustained engraftment and tissue enzyme activity after liver cell transplantation for argininosuccinate lyase deficiency. *Gastroenterology* 130(4), 1317–1323.

Chapter 8

Hepatocyte Transplantation Techniques: Large Animal Models

Anne Weber, Marie-Thérèse Groyer-Picard, and Ibrahim Dagher

Abstract

The poor hepatocyte engraftment efficiency and the low level of their expansion in the host liver are a major limitation to cell therapy for the treatment of life-threatening liver diseases. Many rodent models have shown that liver repopulation via transplanted hepatocytes occurs only when liver growth capacity is impaired for an extended period of time. However, these models are not transposable to the clinics and to date there is no safe method to achieve this result in a clinical setting.

Therefore, it is necessary to define on large animal models strategies that provide to transplanted hepatocytes sufficient proliferation stimuli to induce their division and that could permit a direct extrapolation to humans. Such procedures should be transposable to patients. We have defined a protocol of liver partial portal branch embolisation and shown that it induces the proliferation of transplanted hepatocytes in non-human primates (*Macaca mulatta*). This animal model is also appropriate to evaluate the lentiviral-mediated ex vivo gene therapy approach, since simian hepatocytes are efficiently transduced by HIV-1-derived lentivirus vectors.

Key words: hepatocytes, transplantation, portal embolisation, non-human primates, retroviral transduction.

1. Introduction

The selective replacement of dysfunctional hepatocytes by transplantation of normal hepatocytes has become an alternative to orthotopic liver transplantation for the treatment of life-threatening metabolic diseases and several trials of allogeneic transplantation have already been performed. The overall results suggest that an insufficient number of functional hepatocytes engraft in the liver parenchyma *(1)*. The loss of transplanted hepatocytes prior to their engraftment within the recipient liver parenchyma was also observed in

Anil Dhawan, Robin D. Hughes (eds.), *Hepatocyte Transplantation, vol. 481*

DOI 10.1007/978-1-59745-201-4_8 Springerprotocols.com

non-human primates *(2)*. In parallel, studies in rodents with acute or chronic liver injury showed that transplanted hepatocytes can repopulate recipient livers only when they display a selective advantage over host cells and can proliferate in response to appropriate stimuli *(3–5)*. However, these models are not transposable to the clinics. It is therefore necessary to develop clinically relevant approaches in large animal models, rabbits, pigs, dogs or non-human primates, to increase cell engraftment and proliferation, a limiting step common to allo- and auto-transplantation. Ex vivo gene therapy with autologous hepatocytes would avoid problems related to immunosuppression and the shortage of donor organs. This approach also requires a careful evaluation of transgene expression at long term in situ in such animal models and of its biodistribution.

In humans, partial occlusion either by portal branch ligation or by portal embolisation is currently performed to induce liver regeneration in non-occluded lobes *(6)*. In rats and rabbits, partial portal branch ligation, improves hepatocytes transplantation *(7, 8)*. This procedure developed in *Macaca mulatta* enhances transplanted hepatocyte engraftment *(9)*. Human immunodeficiency virus (HIV)-1-derived vectors transduce efficiently quiescent primary cell types including primary hepatocytes *(10, 11)*. Non-human primate is thus an appropriate model to assay for the long-term expression of therapeutic transgene in situ.

2. Materials

2.1. Animals

Monkeys are *Macaca mulatta*, weighing 3–5.5 kg, seronegative for simian herpes virus, simian retrovirus, simian immunodeficiency virus and simian T-cell lymphotropic virus. All experiments were carried out in accordance with the guidelines of French Ministry of Agriculture.

2.2. Simian Hepatocyte Isolation

1. *Pre-perfusion solution*: 0.1 M Hepes (Free Acid, ULTROL Grade, Merck KGaA, Germany), 0.002 M KCl (Sigma), 0.013 M fructose (Sigma), 0.12 M NaCl (Sigma), 2.8 mM Na_2HPO_4 12 H_2O (Sigma).
2. *Collagenase solution*: Pre-perfusion solution supplemented with 10 mM $CaCl_2$ (Sigma) and collagenase: Worthington type 1 CLS-1 (129 U/ml).
3. *Wash and plating medium*: Dulbecco's Modified Eagle's Medium DMEM/HAMF12 (Eurobio, Les Ulis, France) supplemented with 10% heat-inactivated foetal calf serum (FCS; PAA Laboratories GmbH, Austria), 0.1% bovine serum

albumin, 2 mM L-glutamine and 1% antibiotics (penicillin/streptomycin, 50,000 UI, Eurobio).

2.3. Hepatocyte Culture in Hormonally Defined Medium

DMEM/HAMF12 supplemented with: 1:250 linoleic acid/albumin (Sigma), 5×10^{-8} M 3,3′,5-triiodo-L-thyronine (Sigma), 0.2 IU insulin (Actrapid, Novo Nordisk A/S), 10^{-6} M hydocortisone (Merck Sharp & Dohme), vitamin C (Aguettant, Lyon, France), 0.0025% (w/v) human Apo-Transferrin (iron-poor) (Sigma), 1 mM Na Pyruvate (Eurobio), 2 mM L-glutamine and 1% antibiotics.

2.4. Percoll Solution

To 27 ml Percoll™ (Amersham Biosciences) add 3 ml 10× phosphate-buffered saline (PBS) (Eurobio) and 20 ml plating medium into a 50-ml conical tube. Mix gently upside down several times.

2.5. B-galactosidase Activity

1. Formaldehyde: prepare a 4% solution in PBS fresh for each experiment.
2. Stock solutions: K Ferricyanide: 200 mM in PBS; K Ferrocyanide :200 mM in PBS; $MgCl_2$: 2 M in PBS and substrate X-Gal (5-bromo-4-chloro-3-indolyl-β-D-galactopyranoside): 40 mg/ml in DMSO (stored at –20°C).

2.6. Immuno histochemistry for Green Fluorescent Protein Expression

1. Phosphate-buffered saline (PBS): From 10× stock solution at pH 7.4, prepare working solution by dilution of one part with nine parts water.
2. Formaldehyde (Sigma): Prepare a 4% (v/v) solution fresh for each experiment.
3. Inhibition of endogenous peroxidase solution: 3% H_2O_2 in distilled water.
4. Quench solution: 50 mM NH_4Cl in PBS.
5. Permeabilisation solution: 0.1% (v/v) Triton X-100 in PBS.
6. Blocking solution: 3% (w/v) BSA in PBS.
7. Primary antibody: Anti-GFP antibody, BD Living Colors A.v (Clontech, BD Biosciences, CA, USA).
8. Antibody dilution: 0.1% Tween 20 + 3% BSA in PBS.
9. Secondary antibody: Biotinylated anti-mouse IgG (MOM Vector immunodetection Kit; Vector Laboratories, UK)
10. Covalent conjugate between avidin and an enzyme: peroxidase-conjugated avidin (Vector Laboratories).
11. Peroxidase substrate solution: Diaminobenzidine (DAB) chromogene (Dako K3465)

2.7. BrdU-Labelled Cell Analysis

1. Antigen unmasking solution: Citric acid-based stock solution (Vector, H-3300).
2. ADN denaturation solution: HCl 4 N in water.

3. Washing solution: 0.5% (v/v) Tween in PBS.
4. Inhibition of endogenous peroxidase solution: 5% H_2O_2 in distilled water
5. Primary antibody: mouse monoclonal anti-BrdU antibody: clone Bu20a isotype IgG1k (MO 823) (Dako).
6. Secondary antibody: Biotinylated anti-mouse IgG (Dako, StreptABComplex/HRP Duet Mouse/Rabbit KO492).
7. Covalent peroxidase-conjugated avidin (Dako, StreptABComplex/HRP Duet Mouse/Rabbit KO492).
8. Peroxidase substrate solution: DAB Ultratech, Becton Coulter (IM2394).

3. Methods

Non-human primates are the most closely related to humans. This is true for liver anatomy and hepatic vascularisation, which are different in both dogs and pigs.

Different procedures have been tested on monkeys to partially occlude portal veins. The most efficient one proved to be embolisation with a biological glue, histoacryl, currently used for patients.

Recombinant vectors derived from the onco-retrovirus (Moloney murine leukaemia virus) can be used for gene marking to trace transplanted in situ *(12)*. However, they efficiently transduce only dividing cells and hepatocytes have to be stimulated to proliferate in culture. Lentiviral-mediated transduction of hepatocyte does not require cell division and human immunodeficiency virus (HIV)-1-derived vectors transduce efficiently human and simian hepatocytes. Moreover, hepatocytes can be transduced in suspension immediately after isolation or thawing, which avoids culture and harvest steps *(13)*.

3.1. Removal of the Macaca Left Lobe

1. Operative procedures are performed under general anaesthesia. Monkeys are sedated with an intramuscular injection of ketamine (10 mg/kg intramuscular) and general anaesthesia is induced by the intravenous administration of propofol (2 mg/kg; Diprivan®, Astra-Zeneca, Sodertalje, Sweden) and sufentanil (0.15–0.3 mg/kg, Sufenta®, Janssen-Cilag, Issy-les-Moulineaux, France). Acetaminophen is generally used for analgesia (10 mg/kg orally every 6 h for 3 days).
2. A supraumbilical midline incision is performed. The left lateral lobe is separated from the rest of the liver and removed by cutting the portal pedicle and the corresponding hepatic vein. Haemostasis is achieved by ligature with a 4/0 monoligament thread.

3. Simian hepatocytes are isolated from the left lateral lobe because this lobe is separated from the rest of the liver by a deep fissure and is connected to it only by a narrow parenchymal bridge containing the portal pedicle and hepatic vein. It accounts for about 20% of the liver mass of the cynomolgus monkey *(14)*.

3.2. Portal Embolisation

1. The inferior mesenteric vein is dissected and a 3-F introducer is inserted. An initial portogram is taken to map the portal branches before embolisation. A 3-F angiographic microcatheter (Terumo Progreat® MC-PP27131, Guyancourt, France) is pushed through the portal vein distally into the left and then the right anterior branches.
2. The embolising material (a 1:1 mixture of cyanoacrylate and lipiodol) is injected until complete obstruction of these branches is achieved. Another portogram is then performed to ensure the complete embolisation and patency of the remaining portal branches.
3. The introducer is then replaced by the 4.5-F venous catheter, which is placed right at the junction of the inferior mesenteric vein and the splenic vein. The proximal part of the catheter is connected to a perfusion chamber (Set Celsite® Epoxy Pur 4.5 F, B. Braun Medical, Boulogne-Billancourt, France), placed subcutaneously in the left anterior thoracic region to make repeated access to the portal vein possible.

3.3. Hepatocyte Isolation of Macaca mulatta *Liver*

1. Short plastic catheters (0.7–1.0 mm Vygon, Ecouen, France) are introduced into one (or two) hepatic veins of the resected lobe and secured by a 4/0 ligature and filled with pre-perfusion buffer.
2. A Masterflex Precision tubing (diameter 16 mm) is connected to the catheter introduced in the hepatic vein via a polyethylene extension tube (Vygon) and a double male connector (Vygon).
3. The liver is perfused with 1 l of Hepes buffer pH 7.65 incubated in a water bath at 39°C. The flow rate used varies according to the size of the liver lobe, generally 80 ml/min (*see* **Note 1**).
4. After washing out the blood completely from the liver, it is perfused with 500 ml of Hepes buffer containing 250 mg/500 ml collagenase Type 1 (250 U/mg) (Worthington) supplemented with 10 mM $CaCl_2$ (Sigma) at a flow rate of half of that of the first perfusion. (*see* **Note 2**).
5. The digested liver is transferred into a sterile beaker and 100 ml medium is added. The liver is cut in slices with a scalpel and shaken to release dissociated cells from the Glisson capsula. (*see* **Note 3**).
6. The cell suspension is filtered through sterile gaze to remove small pieces of non-digested liver and transferred into eight

50 ml conical tubes (Falcon). Each tube is adjusted to 50 ml with plating medium.

7. The cells are washed by four centrifugations at $50\times g$ for 3 min at room temperature. After each centrifugation, the supernatant is discarded and the cell pellet gently dissociated in fresh medium. Before the last centrifugation, the cells from four tubes are suspended in 50 ml, and the cell suspension from the four remaining tubes is filtered through a 70-μm nylon filter net into a new sterile bottle, gently mixed and distributed into two 50 ml tubes so that the cell concentration is equal in both tubes. After the last centrifugation, the cells are suspended in 50 ml and counted.
8. The viable cells are counted by dilution of the cell suspension (1:10) into trypan blue solution (0.04% Sigma). Cells with trypan blue-negative nuclei are the viable cells. A Malassez's cell is used to count the cells and calculate the cellular concentration using the formula as follows: Number of viable cells per ml=**n** (number of cell counted)×**f** (dilution factor=10^5 if dilution: 1:10)
9. Hepatocytes are seeded on Primaria culture dishes (Becton Dickinson, USA) in the same plating medium at 2×10^6 cells per 60 mm dish (confluency).
10. The medium is replaced with serum-free medium (HDM) after 5 h, and daily thereafter.

3.4. Percoll Purification

When the recovery of viable cells is less than 85%, it is necessary to perform a Percoll gradient to remove dead cells and cell debris.

For 200 million cells:

1. Twenty-five millilitre of 60% Percoll solution is pipetted into a 50-ml conical tube.
2. Twenty-five millilitre of cell suspension is poured onto the Percoll solution and gently mixed (upside down several times).
3. Hepatocytes are centrifuged at $50\times g$ for 15 min at room temperature.
4. The supernatant is discarded and 40 ml of plating medium are added into each tube. The cell pellet is dissociated by gentle pipetting and centrifuged at $50\times g$ for 5 min. The procedure is repeated twice.
5. The number of viable cells is counted.

3.5. Hepatocyte Labelling with Hoescht Fluorescent Dye

After isolation and eventually Percoll purification, isolated hepatocytes are immediately labelled with the Hoescht fluorescent dye.

1. Hepatocyte suspension is adjusted to 10^7 cells/ml in serum-free medium.

2. One millilitre of hepatocyte suspension is distributed into each 12 ml conical tube.
3. Five microlitres of Hoescht dye is added to the cell suspension, which is incubated for 30 min at 37°C with gentle agitation.
4. The reaction is stopped by the addition of 1 ml FCS and then by the addition of 9 ml medium containing 10% FCS.
5. The cells are centrifuged at $50\times g$ for 5 min, the supernatant is discarded and fresh medium containing 10% FCS is added. The cells are washed three times.
6. Hoescht-labelled hepatocytes are counted and are suspended in plating medium and seeded on culture dishes. Alternatively, hepatocytes are suspended in serum-free medium without phenol red, washed once and suspended in the same medium containing heparin (25 IU/ml) to be infused through the Baby Port.

3.6. Hepatocyte Culture and Retroviral Transduction

Hepatocytes have to be stimulated to proliferate to be transduced by retroviral vectors. This is achieved by the sequential addition of HGF (kindly provided by Genentech, San Francisco, USA) in the HDM medium.

1. The amphotropic FLYTA7 cell line (a gift from F.L. Cosset Inserm France) is used to produce the recombinant retrovirus expressing the β-galactosidase gene under the control of the virus long terminal repeat *(15)*.
2. The cell line is grown in DMEM supplemented with 10^{-3} M sodium pyruvate, 2×10^{-3} M glutamine and antibiotics (Eurobio), and with 10% heat-inactivated FCS.
3. Virus-containing medium is prepared as follows: the night before collection, the medium from confluent plates is removed and replaced with a 1:1 mixture of producer cell medium and hepatocyte medium. The supernatant is harvested 24 h later, filtered through a 0.45-μm pore size filter and immediately frozen in liquid nitrogen and stored at −80°C.
4. Hepatocytes are seeded at 50% confluency (3.5×10^{6} cells) on 100 mm dishes. Hepatocyte growth factor (HGF) is added to the hepatocyte culture 30 h after seeding. Forty-eight hours after seeding, the medium is removed and the plates incubated for 2 h with 500 μl of thawed virus supernatant plus Polybrene (3 μg/ml) (Sigma-Aldrich Co.) in 3 ml medium. HGF is added 4 h before the infection (5 ng/ml).
5. The virus supernatant is then replaced by fresh hepatocyte HDM containing 10 ng/ml HGF.

6. A second infection is performed on day 3. HGF (10 ng/ml) is also added prior to infection and after removal of viral supernatant. On day 4, hepatocytes must reach confluency. (*see* **Note 4**).
7. Summary of the simian hepatocyte transduction: hepatocyte plating density: 3.5×10^6 cells per 100 mm plate; addition of HGF on day 1, twice on day 2 and on day 3; infection 48 and 66 h after plating for 2 h; virus titer > 5×10^7 blue colony-forming unit per millilitre, i.e. multiplicity of infection of 10 (ratio of the number of viral particles to the number of hepatocytes in the dish).
8. Four days after isolation, hepatocytes are stained for β-galactosidase activity or harvested for transplantation *(12)*.

3.7. Lentiviral Transduction of Simian Hepatocytes

The lentiviral vectors are derived from lentivectors of the third generation. They express the green fluorescent protein (GFP) under the control of an endogenous promoter (EF1alpha) and they are produced by Vectalys (Labège, France).

1. Freshly isolated hepatocytes are suspended at 10^6 cells/ml in University of Wisconsin medium containing 50 μM vitamin E (Sigma).
2. Hepatocytes are incubated with lentiviral particles at a multiplicity of infection of 30 for 2 h at 37°C in low attachment plates.
3. The cells are washed five times in plating medium by centrifugation at $50\times g$ for 5 min and then plated on Primaria dishes or transplanted into mouse livers.
4. The cells are cultured during 7 days and then GFP expression is analysed under a fluorescence microscope.
5. Alternatively, cells are harvested for flow cytometer analysis: hepatocytes are incubated for 5 min at 37°C with 2 ml trypsin/10 cm dish (Sigma, T4549). Trypsin activity is then inhibited by the addition of 8 ml plating medium. Hepatocytes are suspended as single cells and centrifuged for 5 min at $50\times g$. Cells are then washed in PBS. After centrifugation, cells are suspended in formaldehyde 1%: 300 μl/10^5 cells and stored at +4°C for cytometer analysis.

3.8. Hepatocyte Transplantation

1. Hoechst-labelled cells are suspended in DMEM medium without phenol red and centrifuged three times at $50\times g$. Extensive washings are necessary to avoid vasoactive shock episodes due to the components of the medium including FCS.
2. Alternatively, 4 days after isolation and retroviral transduction, hepatocytes are harvested with a mixture of 1 ml of 2×10^2 M EDTA in PBS, plus 10 μl of trypsin (Sigma) in

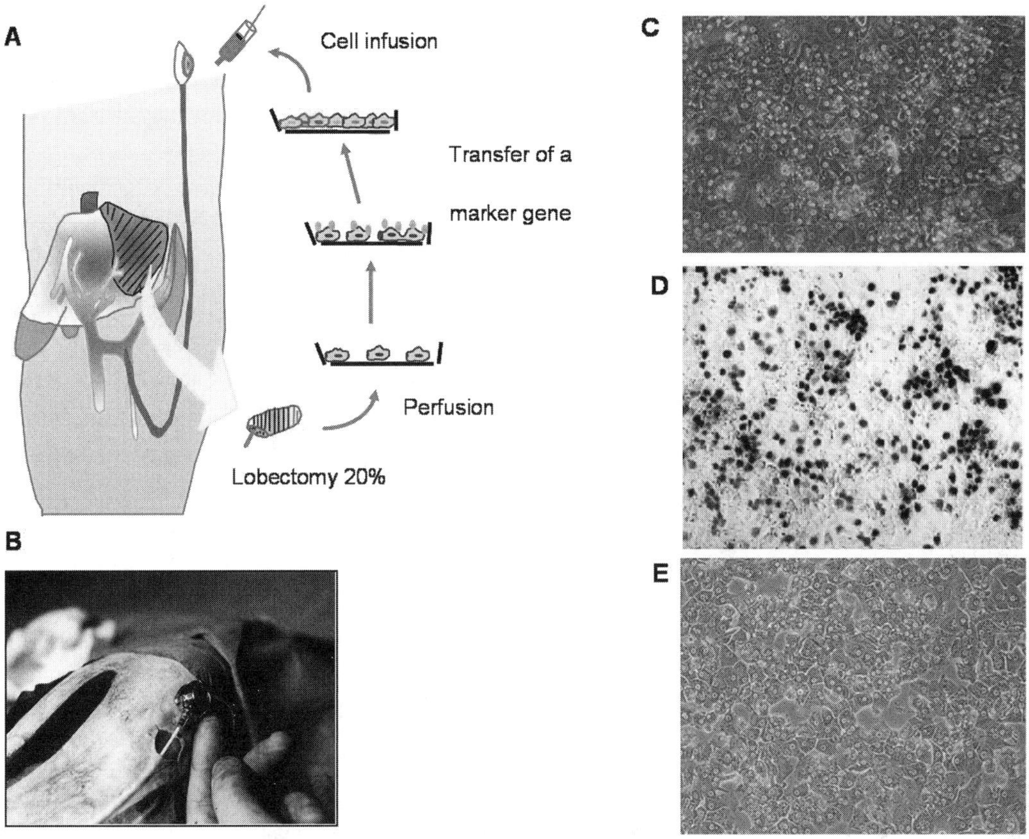

Fig. 8.1. Transplantation of autologous hepatocytes into *Macaca mulatta* after retroviral-mediated gene marking. (**A**) Protocol for simian hepatocyte isolation, retroviral transduction and transplantation. Hepatocyte transduction with HIV-1-derived lentivirus vectors avoids the culture steps. They are transduced in suspension and transplanted. (**B**) Hepatocytes are transplanted via the infusion chamber. (**C**) Freshly isolated simian hepatocytes at confluency after 3 days of culture. (**D**) Transduced hepatocytes in culture expressing the β-galactosidase. (**E**) Thawed hepatocytes after 3 days of culture. (*see* Color Plate 3)

Versene buffer (Gibco/BRL, Bethesda, MD, USA) per 100 mm dish. The cells are suspended into medium containing 2% FCS and washed twice by centrifugation at $50 \times g$ for 5 min, then in serum-free medium without phenol red.

3. The cells are suspended in serum-free medium containing heparin (25 U/ml) (Choay) at a density of 10×10^6 cells/ml and infused through the heparinised Baby Port at a flow rate of 2 ml/min (**Fig. 8.1**).
4. Portal pressure is monitored throughout hepatocyte infusion.
5. Surgical liver biopsies are performed under general anaesthesia at different times after hepatocyte transplantation with a large sample of tissue removed on the edge of each remnant liver lobe through the same midline laparotomy.

6. The liver biopsies are embedded in OCT (Agar), frozen in liquid nitrogen vapours and stored at –80°C. Cryostat sections of 7 μm are performed with cryoultratome (Leica) and examined under fluorescence microscopy (Leica DMR) (excitation at 450 nm) to detect Hoechst-labelled cells.
7. Twenty fields are counted on 10 sections/lobe at ×20 magnification to evaluate the proportion of Hoechst-labelled hepatocytes, knowing that there are 178 hepatocytes in a microscope field.

3.9. Cryopreservation

1. Simian hepatocytes are suspended at a concentration of 5×10^6 cells/ml in plating medium supplemented with 60 μM ZVAD-fmk, Caspase inhibitor (R&D Systems, Minneapolis, MN, USA) and 50 μM vitamin E (Sigma).
2. The cell suspension is incubated for 30 min at 37°C.
3. DMSO (Sigma) is added dropwise and with gentle mixing to give a final concentration of 10%.
4. Hepatocyte suspension is distributed into cryotubes (1 ml/vial), kept for 5 min on ice, then for 2 h at –20°C with upside-down mixing three times every 2 min, then placed overnight (18 h) at –80°C.
5. The following day, the vials are stored in liquid nitrogen.
6. Frozen hepatocytes are thawed by placing the vials directly into a water bath at 37°C.
7. As soon as cells are thawed they are suspended in plating medium in 12 ml conical tubes and centrifuged for 5 min at 50 G.
8. Viable hepatocytes are counted and seeded on collagen 1-coated dishes (BD Bioscience).

3.10. Histochemistry for Detection of β-Galactosidase Activity

1. The hepatocytes are rinsed three times with PBS.
2. Formaldehyde solution is added for 5 min at room temperature to fix the cells, which are then rinsed three times for 10 min each with PBS.
3. Cells are incubated from a few hours to overnight at 30°C in the revealing solution: for 1 ml: 20 μl K ferricyanide; 20 μl K ferrocyanide; 2 μl $MgCl_2$ and 10 μl X-Gal in PBS. (*see* **Note 5**).
4. Cells are then rinsed in PBS and kept in PBS at 4°C. Blue transduced cells are counted under a microscope.

3.11. Immuno histochemistry for Localisation of Transplanted GFP-Expressing Hepatocytes

Several chromogens are used to localise peroxidase in tissue sections. One of the most commonly used has been DAB tetrahydrochloride.

1. Formaldehyde solution is added for 10 min at room temperature to fix the samples.
2. The formaldehyde is discarded and the samples washed three times for 5 min each with PBS.

3. Endogenous peroxidases are inhibited with 3% H_2O_2 in PBS for 30 min at room temperature and washed twice with PBS.
4. Residual formaldehyde is quenched by incubation in NH_4Cl for 15 min at room temperature, followed by three washes in PBS.
5. The samples are permeabilised by incubation in PBS/ 0.1%Triton X-100 for 10 min at room temperature and then rinsed three times with PBS.
6. The samples are blocked by incubation in blocking buffer for 1 h at room temperature.
7. The blocking solution is removed and replaced with the anti-GFP monoclonal antibody (1:100) for 1 h at room temperature in a humid chamber.
8. The primary antibody is removed and the samples washed three times for 5 min each with PBS.
9. The secondary biotinylated antibody is applied according to the M.O.M kit staining procedure and then the sections are washed twice in PBS.
10. The Vectastain ABC reagent is prepared and applied as described in the M.O.M. kit. The sections are incubated for 5 min and then washed twice for 5 min each.
11. DAB solution is applied on the sections: development times, controlled under a microscope, vary between 2 and 10 min in the dark.
12. Sections are then washed in distilled water three times for 2 min each.
13. The samples are then ready to be mounted in glycergel (Dako) or glycerol (90% in PBS) if counter-staining is necessary. (*see* **Note 6**).

3.12. Detection of Dividing Hepatocyte In Situ

Cell division is assessed by BrdU incorporation. BrdU (50 mg/ kg) is infused via the Baby Port for 4 h before liver biopsies are carried out.

1. Liver sections are deparaffinised through xylene and graded alcohol series three times for 10 min and rinsed in tap water.
2. The slides are rapidly rinsed in distilled water.
3. Citrate buffer (1:100) is then added and the slides are placed in a microwave oven at 650 W for 5 min and at 160 W for 15 min to unmask the specific antigens and then rinsed twice in distilled water.
4. ADN is denatured with HCl 4 N for 20 min and the sections are rinsed three times with distilled water, then rinsed in 0.5% PBS/Tween twice for 5 min.
5. Endogenous peroxidase activity is inhibited with 5% H_2O_2 for 10 min and then the samples are rinsed with distilled water.

6. The non-specific sites are blocked by incubation in goat serum (1:20) for 10 min at room temperature, then the excess of serum is removed without rinsing.
7. The samples are incubated with anti-BrdU monoclonal antibody (1:100) in antibody dilution buffer for 1 h at room temperature in a humid chamber, then washed three times for 5 min each with PBS.
8. The secondary biotinylated antibody is applied according to an indirect avidin–biotin peroxidase kit for 15 min and then the sections are washed twice in PBS/Tween.
9. The complex strepavidin–peroxidase is added for 15 min (kit Dako) and then the sections are washed twice in PBS/Tween.
10. DAB solution is applied on the sections: development times, controlled under a microscope, 10 min in the dark.
11. Harris hematoxylin solution is applied for 5 min. then the samples are rinsed three times in tap water and in distilled water.
12. The samples are dehydrated in graded alcohol series, then placed in xylene three times for 5 min and then mounted glycergel (Dako).

4. Notes

1. The pre-perfusion has to be flowed until the blood is completely washed out from the liver lobe. Stop the flow before air bubbles move into the liver. The portal vessels allow to flow the perfusate out of the lobe and to avoid an increase in the pressure.
2. The batch of collagenase is critical for cell viability and transduction efficiency. Batches are first tested for their ability to produce high yields, maximum viability and membrane recovery of rat hepatocytes. Currently, collagenase A from Boehringer (Mannheim, Germany) or collagenase type 1 CLS-1 (Worthington) is used. Collagenase must be dissolved when the amount of the pre-perfusion solution becomes small to avoid a decrease in collagenase activity. To preserve the maximum of enzyme activity and to avoid too much cooling of collagenase solution in the tubing, the water bath temperature is kept at 39°C.
3. Liver digestion has to be carefully checked and, depending on lobe size, collagenase perfusion can be stopped before the end of the solution flows out.
4. A low number of hepatocytes per dish leads to their apoptosis. The number of hepatocytes should be carefully adjusted to 50% confluency when retroviral transduction is performed.

The plating efficiency is always inferior to the number of viable cells as assessed by trypan blue.

5. To detect β-galactosidase activity, culture dishes or sections have to be incubated at 30°C rather than at 37°C, because at this temperature, endogeneous β-galactosidase is not revealed. Whenever possible, it is recommended to add a nuclear localisation signal (nls) that targets the protein to the outer membrane of the nucleus and distinguish it from the endogenous lysosomal enzyme.
6. Hepatocytes in liver sections are autofluorescent. Therefore, GFP-transduced and GFP-transplanted hepatocytes are generally difficult to detect from the resident cells. It is therefore best to use an anti-GFP antibody to detect the genetically modified engrafted cells.

Acknowledgments

The authors thank Pr Dominique Franco for his permanent support as well as all the members of Inserm U 804 who participated in these protocols. Experiments on animals were performed at INRA (Jouy-en-Josas), and we thank Dr Guy Germain and Dr Alexandre Laurent for their help and advice.

This work was supported by AFM (Association Française contre les Myopathies), Inserm, University Paris XI, Délégation à la Recherche Clinique AP-HP.

References

1. Fisher, R. A., Strom, S. C. (2006) Human hepatocyte transplantation: worldwide results. *Transplantation* 82, 441–449.
2. Weber, A., Mahieu-Caputo, D., Hadchouel, M., et al. (2006) Hepatocyte transplantation: studies in preclinical models. *J Inherit Metab Dis* 29, 436–441.
3. Allen, K., Soriano, E. (2001) Liver cell transplantation: the road to clinical application. *J Lab Clin Med* 138, 298–311.
4. Grompe, M. (2006) Principles of therapeutic liver repopulation. *J Inherit Metab Dis* 29, 421–425.
5. Azuma, H., Paulk, N., Ranade, A., et al. (2007) Robust expansion of human hepato cytes in Fah-/-/Rag2-/-/Il2rg-/- mice. *Nat Biotechnol* 25, 903–910.
6. Makuuchi, M., Thai, B. L., Takayasu, K., et al. (1990) Preoperative portal embolization to increase safety of major hepatectomy for hilar bile duct carcinoma: a preliminary report. *Surgery* 107, 521–527.
7. Ilan, Y., Roy-Chowdhury, N., Prakash, R., et al. (1997) Massive repopulation of rat liver by transplantation of hepatocytes into specific lobes of the liver and ligation of portal vein branches to other lobes. *Transplantation* 64, 8–13.
8. Eguchi, S., Rozga, J., Lebow, L. T., et al. (1996) Treatment of hypercholesterolemia in the Watanabe rabbit using allogeneic hepatocellular transplantation under a regeneration stimulus. *Transplantation* 62, 588–593.
9. Dagher, I., Boudechiche, L., Branger, J., et al. (2006) Efficient hepatocyte engraftment in a nonhuman primate model after partial portal vein embolization. *Transplantation* 82, 1067–1073.

10. Nguyen, T. H., Birraux, J., Wildhaber, B., et al. (2006) Ex vivo lentivirus transduction and immediate transplantation of uncultured hepatocytes for treating hyperbilirubinemic Gunn rat. *Transplantation* 82, 794–803.
11. Nguyen, T. H., Oberholzer, J., Birraux, J., et al. (2002) Highly efficient lentiviral vector-mediated transduction of nondividing, fully reimplantable primary hepatocytes. *Mol Ther* 6, 199–209.
12. Andreoletti, M., Loux, N., Vons, C., et al. (2001) Engraftment of autologous retrovirally transduced hepatocytes after intraportal transplantation into nonhuman primates: implication for ex vivo gene therapy. *Hum Gene Ther* 12, 169–179.
13. Parouchev, A., Nguyen, T. H., Dagher, I., et al. (2006) Efficient ex vivo gene transfer into non-human primate hepatocytes using HIV-1 derived lentiviral vectors. *J Hepatol* 45, 99–107.
14. Vons, C., Loux, N., Simon, L., et al. (2001) Transplantation of hepatocytes in nonhuman primates: a preclinical model for the treatment of hepatic metabolic diseases. *Transplantation* 72, 811.
15. Cosset, F. L., Takeuchi, Y., Weiss, R., et al. (1995) High-titer packaging cells producing recombinant retrovirus resistant to human serum. *J Virol* 69, 7430–7436.

Chapter 9

Cell Transplant Techniques: Engraftment Detection of Cells

Robert A. Fisher and Valeria R. Mas

Abstract

The use of isolated human hepatocyte infusions to treat human disease will require safe, acceptable, reliable, and reproducible measures of engraftment and function of the donor liver cell. Cell transplant for inborn errors of hepatic metabolism can be followed by measuring the specific protein missing from the recipient, expressed by the transplanted unmodified donor hepatocytes expressing the genes in question. This chapter will focus on the clinical techniques successful in identifying the engraftment and function of donor human hepatocytes when no specific identifiable genes are expressed by donor hepatocytes in acute and chronic liver diseases treated by cell infusion. Radiolabeling and dye labeling techniques, DNA typing of HLA class I alleles, soluble class I HLA ELISA, real-time quantitative PCR techniques including short tandem repeats analysis will be detailed and critiqued.

Key words: Human hepatocyte, short tandem repeats (STR), SHLA-class I, Real-time PCR.

1. Introduction

The first illustrations published on using a cell labeling technique in human hepatocyte transplantation used 99^{m} Tc (technetium) scintigrams to detect hepatocyte autotransplants, injected into the spleen, detected at 1 and 10 months follow-up *(1)*. The use of radiolabeling technology to follow human allogeneic hepatocyte transplant in the spleen, in the later 1990 s, was demonstrated using serial technetium – 99m-diisopropyl-iminodiacetic acid (DISIDA) serial perfusion scans from days 2 to 23 post cellular infusion. The Tc scans combined with serum measured serial improved ammonia clearance matched radiologic evidence of hepatocellular activity in the spleen with hepatocellular function *(2)*.

Anil Dhawan, Robin D.Hughes (eds.), *Hepatocyte Transplantation, vol. 481*

DOI 10.1007/978-1-59745-201-4_9 Springerprotocols.com

To provide short-term (7 days) noninvasive analysis of the biodistribution of human hepatocytes infused into a 5-year-old with ornithine transcarbamylase (OTC) deficiency, 10^8 donor hepatocytes were radiolabeled using indium-111 oxyquinoline solution.

The use of hepatocyte dye labeling technique in rat and porcine hepatocytes, using carboxyfluorescein (CFSE) and DiL have provided elegant data on the number and location of engrafted hepatocytes in animal studies of cellular transplantation *(3)*. These dye techniques, to our knowledge, have not been duplicated in human cell transplant studies.

The novel idea that HLA class I tissue typing together with serial ELISA measurement of (soluble) sHLA class I antigen could be a practical, safe, and specific method of following donor hepatocyte engraftment into recipient liver with a genetically different class I HLA was based on the routine availability of tissue typing expertise at transplant centers, and the knowledge that all liver allografts produce sHLA-I Ag within minutes of implantation and maintain high and stable sHLA-I Ag release with stable liver allograft function *(4)*. Furthermore, the prospective measurement of HLA class I Ag as a marker of donor hepatocyte viable engraftment was chosen over HLA class II Ag, because the accuracy of ELISA in correlating light absorbance to pure standard controls of HLA-I are stable and more consistent than sHLA-II; and unlike sHLA-II, sHLA-I Ag secretion relationship to allotypes in human populations has been studied and confirmed *(4, 5)*.

Hepatocyte engraftment in a human liver with one cell infusion is typically lower than 1% of the total liver mass. Real-time PCR techniques have been developed with sensitivities as low as 0.01% to assess minute levels of repopulation and chimerism. The majority of these published applications have studied liver tissue after sex-mismatched hepatocyte transplantation by real-time quantitative PCR for Y chromosome sequences, not helpful in sex-matched liver cell transplantation, thus limiting the broad clinical application *(6, 7)*.

Short tandem repeats (STR) are highly polymorphic DNA sequences in the human genome used as a standard tool for human identity testing *(8, 9)*. Because of their high level of polymorphism, combined with the simplicity of their analysis, these markers are appropriate by engraftment studies. Coupling PCR to the use of a fluorescence DNA analyzer permits accurate measurement of the amount of PCR product and development of quantitative assays. A sensitive, simple, and specific method of monitoring the engraftment of transplanted hepatocytes using STRs combined with a repeatable, reliable technique for using paraffin-embedded tissue specimens is described *(10)*.

2. Materials

2.1. Cell Labeling with Indium

1. Indium-111 oxyquinoline solution (1 mCi/ml activity; Amersham Corp., Arlington Heights, IL, USA).

2.2. HLA Class I Tissue and Soluble Typing

1. Mouse anti-human monoclonal antibodies (One Lambda Inc. Canoga Park, CA, USA) microtiter plates (CoStar, Cambridge, MA, USA).
2. Rabbit anti-human β-2 microglobulin (Dako, Carpinteria, CA, USA).
3. Tetramethylbenzidine (Dako), which is the substrate of peroxidase.

2.3. Real-Time Quantitative PCR and STR Techniques

1. QIAamp Tissue Kit (Qiagen, Valencia, CA, USA).
2. PCR-SSP typing tray was from One Lambda Inc.
3. Perkins-Elmer Ampli *Taq* DNA polymerase (Norwalk, CT, USA).
4. PE 9700 Thermocycler (Perkin-Elmer).
5. Agarose gel with Micro SSP Gel System (One Lambda Inc.).
6. The AmpFLSTR® Profiler Plus™ PCR Amplification Kit (Applied Biosystems, Foster City, CA, USA).
7. 310 Genetic Analyzer (Applied Biosystems).

3. Methods

3.1. Cell Labeling with Indium

1. The procedure in brief is 10^8 human hepatocytes suspended in serum-free phosphate-buffered saline (PBS), centrifuged at $70\times g$ for 10 min.
2. The cells are then re-suspended with (1.3 mCi) In-111 oxyquinoline drop by drop with gentle shaking. The suspension is gently agitated for 20 min of incubation at room temperature.
3. The In – 111 hepatocytes are re-suspended twice in 10 ml ice-cold PBS, and centrifuged twice at $70\times g$ for 10 min, each time. The re-suspension and centrifugation is repeated for every 30 min storage interval, until patient infusion, to ensure the complete removal of unbound radioactivity (*see* **Note 1**). This procedure provides a labeling cell efficiency of 36%, which is adequate for clinically useful scintigraphy *(11)*.

3.2. HLA Class I Tissue and Soluble Typing

3.2.1. Class I-Specific ELISA

The methods used for class I-specific ELISA in brief are *(4, 12, 13)*: Mouse anti-human monoclonal antibodies were used to measure donor-specific sHLA. Donor-mismatched HLA alleles were chosen to avoid known cross-reactivity with other recipient HLA alleles.

1. Plasma samples are analyzed at a half dilution and all samples are tested on the same day to minimize interassay variations.
2. Briefly, microtiter plates are coated with 100 μl of the chosen anti-sHLA overnight at 4°C. The two or three chosen antibodies are diluted 1:200 in a carbonate buffer (35 mM $NaHCO_3$/15 mM Na_2CO_3, pH 9.6).
3. Free binding sites are blocked by incubation of 200 μl PBS containing 0.05% Tween 20 (PBST) and 1% bovine serum albumin for 1 h at 37°C.
4. The plasma samples are centrifuged at 14,000×*g* for 5 min to remove undissolved proteins. One hundred microliters of the patient's serum is added in half dilution with PBST and incubated for 2 h at 37°C.
5. Subsequently, 100 μl of rabbit anti-human β-2 microglobulin is added in 1:1000 dilution with PBST and incubated for 1 h at 37°C.
6. Finally, the plate is washed extensively three times with PBST and incubated with 100 μl of conjugated goat anti-rabbit IgG-horseradish peroxidase in 1:5000 dilution at 37°C for 1 h. After the wash with PBST, bound antibody is detected by adding 100 μl of tetramethylbenzidine, which is the substrate of peroxidase.
7. The reaction is stopped after 20 min with 100 μl of 2.5 N H_2SO_4 and the absorbance read at 450 nm. Background control uses PBST containing 1% bovine serum albumin, and the absorbance is subtracted by background reading.

3.2.2. Micro SSP DNA Tissue Typing

1. For DNA extraction from the biopsy tissue, the QIAamp Tissue Kit is used. Briefly, the tissue is cut into small pieces. Proteinase K is used to mix with the tissue at 55°C until the tissue is completely lysed.
2. RNase A (20 mg/ml) is added to digest RNA in the liver tissue.
3. After 100% ethanol precipitation, the samples were placed on a QIAamp spin column and centrifuged at 6000×*g* for 1 min. DNA samples are eluted with distilled water and the concentration of DNA measured *(14)*.
4. The Micro SSP DNA Typing Tray is a polymerase chain reaction sequence-specific primer (PCR-SSP)-based assay for the DNA typing of HLA class I alleles *(15)*. This technique

determines whether donor-specific HLA is present (chimerism) in the pool of liver biopsy specimens from the patient.

5. All procedures were strictly followed according to the manufacturer's instructions. Each run of PCR includes a negative control. The presence of the negative control band and/or the positive typing band in the negative control well voids all test results.
6. The master mix is prepared, and 28 U of Ampli *Taq* DNA polymerase is used for each tray. The tray containing complete reactions is placed on a PE 9700 Thermocycler *(16)*. The PCR program is run as follows: 1 cycle of 96°C for 140 s, 65°C for 60 s; 5 cycles of 96°C for 20 s, 65°C for 60 s; 20 cycles of 96°C for 20 s, 59°C for 30 s, 72°C for 45 s; and 8 cycles of 96°C for 20 s, 55°C for 60 s, 72°C for 90 s.
7. Each result is examined on a 2.5% agarose gel with a Micro SSP Gel System (*see* **Notes 2 and 3**).

3.3. Real-Time Quantitative PCR and STR Techniques

1. The assay characteristics and analytical validation in brief are: The AmpFLSTR® Profiler Plus™ PCR Amplification Kit amplifies nine tetranucleotide STR loci and the amelogenin locus in a single reaction tube. The STR loci amplified are D3S12358, D5S818, D7S820, D8S1179, D18S51, D21S11, FGA, and vWA. The amelogenin locus is used for gender identification because products of different lengths are generated from the X and Y chromosomes (**Fig. 9.1**).
2. Engraftment analysis requires one or more informative loci that distinguish the recipient from the donor. Each selected polymorphism is tested by means of an artificial reconstruction mixture of varying percentages of informative pre-transplant recipient and donor DNAs to determine the validity and the sensitivity of the method.
3. Using 11 dilutions simulates a range of mixed chimerisms varying from 100 to 0.01% (90, 70, 50, 25, 10, 5, 1, 0.75, 0.5, 0.1, and 0.01%).
4. In addition, a negative control (100% donor DNA for recipient marker amplification and the converse for donor marker amplification) is included in the assay.
5. Each mix sample dilution is run in triplicate and the complete experiments are run twice on 2 different days and are conducted by the same operator.
6. The mixing of DNAs is conducted on freshly collected human peripheral blood with similar white blood cell counts. In addition, sex-matched and mismatched cases are included for the analytical validation.
7. DNA is isolated from individual blood mixtures.
8. Finally, PCR amplification is performed in triplicate according to the manufacturer's instructions (using 25 cycles) and all

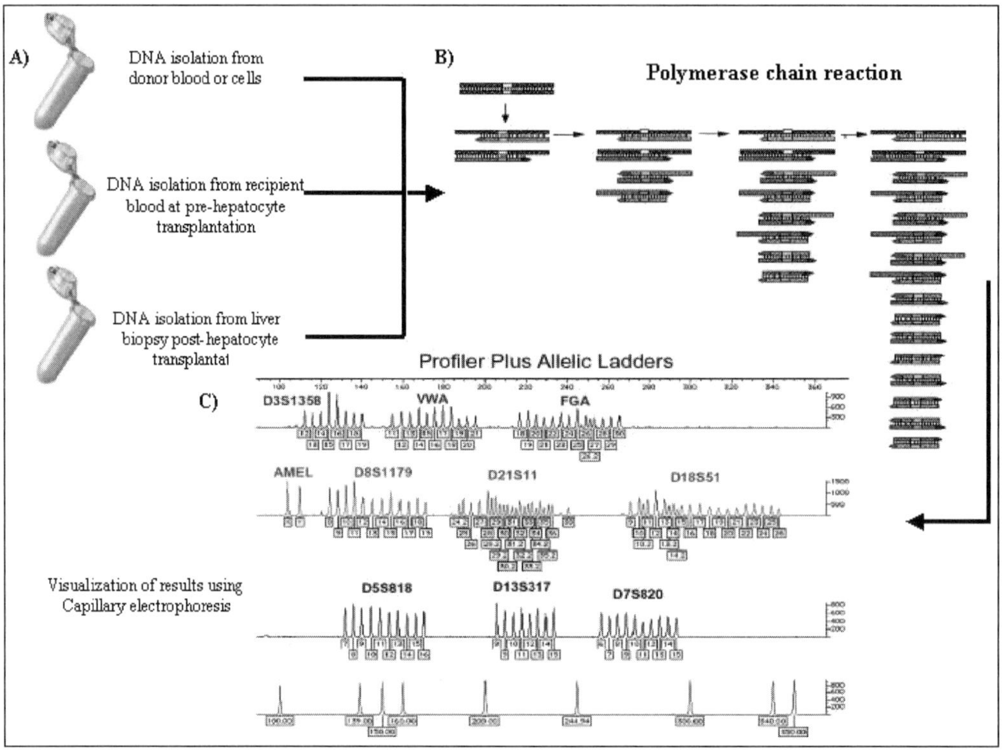

Fig. 9.1. (**A**) DNA is to be isolated from donor and recipient before and after hepatocyte transplantation and then (**B**) amplified to produce sufficient DNA quantity so that (**C**) the AmpFLSTR Profiler Plus PCR Amplification Kit (Applied Biosystems) can be used to quantify the donor-to-recipient DNA ratio to determine the donor cellular engraftment of biopsies of transplanted site or sites at variable times with accuracy, reproducibility, and sensitivity (0.5% donor DNA/ recipient DNA).

the samples are analyzed on a 310 Genetic Analyzer in the same run.

9. In addition, DNA mixes for the sensitivity analysis are created from DNA isolated from paraffin-embedded liver tissues (PELT). The sensitivity of the test was established at 0.5% of DNA donor in the recipient using at least two informative alleles for the final engraftment percentage calculation. Differences in the sensitivity between the curves of DNA mixes from peripheral blood cells and PELT are not observed. Donor genotype is detected until the 0.5% recipient cell fraction with at least two informative markers. Using a linear regression analysis, comparing measured donor genotype (%) versus effective donor DNA (%), the value for the coefficient of determination r^2 was 0.988. (*see* **Notes 4 and 5**).

4. Notes

1. This re-suspension procedure, to minimize radiation injury to labeled cells and provide the minimal cell labeling efficiency, reduces cell viability by as much as 10–20% even with the use of better cell-enhancing supernatants *(17)* and shorter storage time (<2 h) in the authors in vitro experiments to improve In-111 human hepatocyte labeling methods. These factors have limited the routine application of this laborious labeling technique to clinical human cell transplant study.
2. Using the techniques detailed above, the first isolated human liver cell transplant for fulminant liver failure as a bridge to native liver regeneration was safely and reliably verified *(18)*. The limitations of future applications of HLA class-I Ag monitoring for human hepatocyte transplant engraftment were primarily the senior authors inability to finance, test, and maintain a broad enough library of readily available anti-HLA class I monoclonal antibodies to be available for the diverse unpredictable donor cell to recipient HLA class I combinations based on donor hepatocyte availability and affected recipient populations.
3. Finally, the other major critique of this methodology for following human hepatocyte engraftment and viability was that it lacked a simultaneous hepatocellular function-specific quantification that has been solved with the next described applied technology.
4. By combining use of the AmpFLSTR Profiler Plus PCR Amplification Kit and quantification of gene expression of the liver-specific transcripts, albumin, and P450 II B1, using real-time PCR, successful human hepatocyte transplant engraftment in a liver failure patient bridged to native liver regeneration (**Fig. 9.2**) was specifically, quantifiably, and reproducibly measured *(19)*.
5. Finally, by combining STRs with liver function-specific transcripts in a gene array chip platform, we will automate and standardize measurements of hepatocyte engraftment and function. Although cell transplant for non-inborn errors of hepatic metabolism have stimulated these molecular techniques, they will ironically aid in the timing studies of additional cell transplants for the treatment of inherited hepatocellular factor deficiencies that are not solved by single cell infusion and simple-factor (i.e. Factor VII) *(20)* serum measurement.

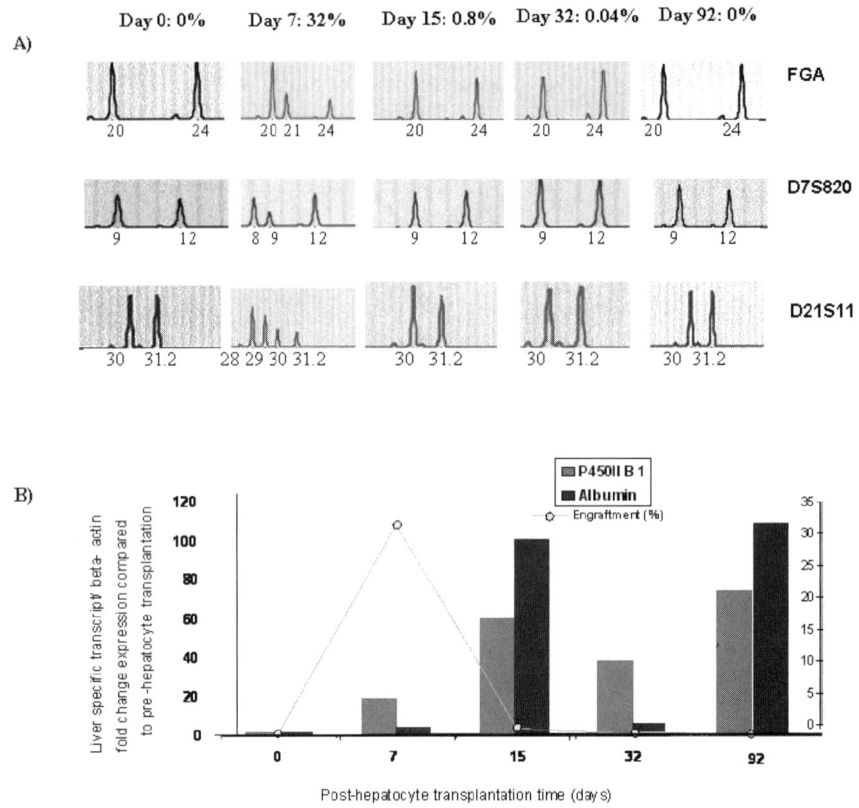

Fig. 9.2. Study of engraftment and gene expression after hepatocyte transplantation. (**A**) The AmpFLSTR Profiler Plus PCR Amplification Kit (Applied Biosystems) is used for the study of engraftment. Engraftment studies are performed in liver biopsies at days 0, 7, 15, and 32 post hepatocyte transplantation. Although the pre-transplantation biopsy shows markers corresponding to the recipient genotype, a mix of markers from donor and recipient is observed at day 7 with lower engraftment percentages at days 15 and 32. (**B**) Quantitation of gene expression of the liver-specific transcripts albumin and P450IIB1 is performed using real-time PCR. MRNA levels of both transcripts are increased after transplantation when compared with pre-transplantation values.

References

1. Kusano, M., Jiang, B., Murakami, M., et al. (1997) Clinical liver cell transplantation, in (Mito, M., Sawa, M., eds.), *Hepatocyte Transplantation: Now and Then,* pp. 297–311. Karger Landes Systems, Basel, Switzerland.
2. Fisher, R. A., Strom, S. C. (2000) Human hepatocyte transplantation: biology and therapy. in (Berry, M. N., Edwards, A. M., eds.), *In the Hepatocyte Review,* pp. 475–501. Kluwer Academic Publishers, Dordrecht, The Netherlands.
3. Fujioka, H., Hunt, P. J, Rozga, J., et al. (1994) Carboxyfluorescein (CFSE) Labeling of hepatocytes for short-term localization following intraportal transplantation. *Cell Transplantation* 3, 397–408.
4. McDonald, J. C., Adamashivili, I. (1998) Soluble HLA: a review of the literature. *Human Immunol* 59, 387–403.
5. McDonald, J. C., Adamashivili, I., Zobaro, G. B., et. al. (1997) Serologic allogeneic chimerism. *Transplantation* 64 (6), 865–871.
6. Byrne, P., Huang, W., Wallace, V. M., et al. (2002) Chimerism analysis in sex-mismatched murine transplantation using guantitative real-time PCR. *Biotechniques* 32, 279–280, 282–284.

7. Wang, L. J., Chen, Y. M., George, D., et al. (2002) Engraftment assessment in human and mouse liver tissue after sex-mismatched liver cell transplantation by real-time quantitative PCR for Y chromosome sequences. *Liver Transpl* 8, 822–828.
8. Antin, J. H., Childs, R., Filipovich, A. H., et al. (2001) Establishment of complete and mixed donor chimerism after allogeneic lymphohematopoietic transplantation: Recommendations from a workshop at the 2001 Tandem Meetings of the International Bone Marrow Transplant Registry and the American Society of Blood and Marrow Transplantation. *Biol Blood Marrow-Transplant* 7, 473–485.
9. Kleeberg, W., Rothamel, T., Glockner, S., et al. (2002) High frequency of epithelial chimerism in liver transplants demonstrated by microdissection and STR-analysis. *Hepatology* 35(1), 110–116.
10. Mas, V. R., Maluf, D. G., Thompson, M., et al. (2004) Engraftment measurement in human liver tissue after liver cell transplantation by short tandem repeats analysis. *Cell Transpl* 13, 231–236.
11. Bohnen, N. I., Charron, M., Reyes, J., et al. (2000) Use of Indium – III – labeled hepatocytes to determine the biodistribution of transplanted hepatocytes through portal vein infusion. Clin Nucl Med 25, 447–450.
12. Koelman, C. A., Mulder, A., Jutte, N. H., et al. (1998) The application of human monoclonal antibodies for monitoring donor derived soluble HLA Class I molecules in the serum of heart transplant recipients. *Human Immunol* 59, 106–114.
13. Pouletty, C., Mercier, I., Glanville, L., et al. (1994) Typing of a panel of soluble HLA class I antigen by enzyme-linked immunosorbent assay. *Human Immunol* 40, 218.
14. Ausubel, F. M., Brent, R., Kingston, R. E., et al. (1992) *Current Protocols in Molecular Biology*. New York: John Wiley and Sons, 59.
15. Teraski, P. I. (1980) Histocompatibility testing. Report of the 8th International histocompatibility workshop, in (Taraski, PI, ed.), UCLA Tissue Typing Laboratory, Los Angeles, CA.
16. Newton, C. R., Graham, A., Heptinstall, E., et al. (1989) Analysis of any point mutation in DNA: the amplification refractory mutation system (ARMS). *Nucleic Acids Res* 17, 2503.
17. Fisher, R. A., Bu, D., Thompson, M., et al. (2004) Optimization of conditions for clinical human hepatocyte infusion. *Cell Transpl* 13, 677–689.
18. Fisher, R. A., Bu, D., Thompson, M., et al. (2000) Defining hepatocellular chimerism in a liver failure patient bridged with hepatocyte infusion. *Transplantation* 69, 303–307.
19. Fisher, R. A., Strom, S. C. (2006) Human hepatocyte transplantation: Worldwide results. *Transplantation* 82, 441–449.
20. Dhawan, A., Mitry, R. R., Hughes, R. D., et al. (2004) Hepatocyte transplantation for inherited factor VII deficiency. *Transplantation* 78, 1812–1813.

Chapter 10

Hepatic Preconditioning for Transplanted Cell Engraftment and Proliferation

Yao-Ming Wu and Sanjeev Gupta

Abstract

Hepatocyte transplantation has therapeutic potential for multiple hepatic and extrahepatic disorders with genetic or acquired basis. To demonstrate whether cell populations of interest will be effective for clinical applications, it is first necessary to characterize their properties in animal systems. Demonstrating the potential of cells to engraft and proliferate is a critical part of this characterization. Similarly, for stem/progenitor cells, demonstrating the capacity to differentiate along appropriate lineages and generate mature cells that can engraft and proliferate is essential. In various animal models, preconditioning of recipients prior to cell transplantation has been necessary to improve engraftment of cells, to stimulate proliferation of engrafted cells, and to induce extensive repopulation of the host liver by transplanted cells. Although this is an area of active investigation, effective preconditioning protocols should alter the hepatic microenvironment, such that transplanted cells can obtain selective advantages for engrafting and proliferating in the liver. Use of such experimental systems in animals will help generate further strategies for liver repopulation and thereby advance clinical applications of liver cell therapy.

Key words: Hepatocyte, engraftment, liver, preconditioning, proliferation, transplantation.

1. Introduction

In principle, liver-directed cell therapy could substitute for orthotopic liver transplantation (OLT) in some conditions, serve as a bridge to OLT in other situations, e.g., acute liver failure, and offer the possibility of hepatic support in refractory hepatic failure, when OLT may not be possible. To advance applications of hepatocyte transplantation, it is necessary to develop effective mechanisms for engraftment and proliferation of transplanted cells, such that the liver can be repopulated to the desired extent. Moreover, it is necessary to identify suitable

Anil Dhawan, Robin D.Hughes (eds.), *Hepatocyte Transplantation, vol. 481*

DOI 10.1007/978-1-59745-201-4_10 Springerprotocols.com

cell populations for transplantation, which will be capable of engrafting, proliferating, and restoring deficient function under various circumstances. These goals require the availability of appropriate systems in vivo using both small and large animals.

Apart from issues concerning the route of cell delivery, infusion rate, prior manipulations of cells, fresh versus frozen cells, etc., the successful engraftment of cells constitutes the first step of effective cell transplantation. In general, transplantation in one session of the equivalent of 2–5% of the hepatocyte mass present in the whole liver is well tolerated and is without serious adverse effects. However, only 15–20% of transplanted hepatocytes engraft successfully in the parenchyma of the recipient liver *(1, 2)*. Therefore, not more than 1% of the liver can be replaced by transplanted hepatocytes after one session of cell transplantation. Repeated cell transplantation can increase the fraction of transplanted cells in the liver, although the extent of liver replacement remains limited (5–7%). On the other hand, effective cell therapy demands greater liver replacement. Therefore, efforts have been ongoing to develop suitable strategies for obtaining superior results following hepatocyte transplantation.

Two complementary approaches have been effective in increasing the number of transplanted hepatocytes in the liver. The first approach concerns the improvement of cell engraftment by various manipulations, including repeated cell transplantation, use of vasodilators to alter the distribution of transplanted cells in the liver lobule, inhibition of macrophage function, manipulation of extracellular matrix component interactions in liver sinusoids, and prior disruption of the sinusoidal endothelial barrier with specific drugs or chemicals *(3–8)*. The second approach concerns the induction of proliferation in transplanted hepatocytes to promote repopulation of the recipient liver. Creation of a suitable hepatic microenvironment may stimulate proliferation in transplanted cells, although when regenerating native cells compete with transplanted cells, e.g., in response to partial hepatectomy or ischemic liver injury, transplanted cells do not proliferate beyond 1–2 cell doublings. The most effective way to induce proliferation in transplanted cells is to either selectively enhance the proliferation capacity of transplanted cells *(9)* or to impair survival and/or proliferation in native cells *(10–17)*. The former approach requires manipulation of cell cycle regulatory controls, which is intrinsically problematic due to the possibility of oncogenic perturbations. Therefore, recent interest has focused more on the latter approach.

The combination of manipulations to improve cell engraftment in the first instance followed by perturbation of native hepatocytes to induce proliferation in transplanted cells has been most effective. It should be noteworthy that manipulations capable of improving cell engraftment can increase the number of engrafting cells by several-fold, which can greatly accelerate the kinetics of liver repopulation. The time taken to near-total liver repopulation under such circumstances can be markedly shortened. The findings are significant because transplanted hepatocytes survive life-long in the absence of rejection and this has obvious implications for therapies in specific disorders.

Several preconditioning regimens were recently developed to improve engraftment and proliferation of transplanted cells in animals. Rodent models capable of demonstrating transplanted cell proliferation include transgenic strain combinations, e.g., alb-uPA transgenic mice as recipients, Bcl2–/– mice as donors with Jo-2 Fas-ligand-induced liver damage in recipients, mice lacking the fumarylacetoacetate hydroxylase enzyme (FAH–) as recipients *(10–12)*, use of radiation plus hepatic ischemia and reperfusion *(13)*, radiation plus hepatocyte growth factor *(14)*, retrorsine plus partial hepatectomy or thyroid hormone or carbon tetrachloride *(15–18)*, and use of monocrotaline plus partial hepatectomy or CCl4 *(19, 20)*. Among these, protocols using hepatic radiation could be useful for clinical applications and these are undergoing further investigations.

On the other hand, convenient protocols for animal studies will be particularly helpful in preclinical studies, analysis of lot-to-lot variability of cells during clinical trials, as well as assessment of novel cell populations. Derivatives of embryonic stem cells, organ-derived stem/progenitor cells, and circulating stem/progenitor cells gained interest for cell therapy, although understanding their properties in vivo requires further work. Also, suitable animal models will help in mechanisms concerning xenotransplantation of cells, e.g., use of porcine hepatocytes, which show therapeutic potential in rodent systems, although further analysis is required on how these cells will engraft, proliferate, and function in the liver.

Here, we provide convenient protocols in rats to establish mechanisms in the engraftment and proliferation of transplanted hepatocytes. The preconditioning regimens described utilize the pyrrolizidine alkaloids, retrorsine and monocrotaline, which exert hepatic toxicity and are known to possess oncogenic potential. Therefore, these chemicals are not suitable for clinical use.

2. Materials

2.1. Animals

1. Six to 10-week-old dipeptidyl peptidase-deficient (DPPIV–) rats in F344 background as cell recipients (F344/DchcHsd-*DPPIV*–; Harlan Sprague Dawley Inc., Indianapolis, IN, USA).
2. Inbred F344 rats as hepatocyte donors (F344/NHsd from Harlan Sprague Dawley Inc.).

2.2. Hepatocyte Isolation (see previous detailed publication in Ref. (21)

1. Perfusion solution:(a) Leffert's buffer, (b) EGTA in Leffert's buffer, and (c) collagenase *(21)* (all chemicals from Sigma Chemical Co., St. Louis, MO, USA).
2. Nylon mesh 85 μm pore size.
3. RPMI 1640 medium.
4. Trypan blue dye 0.4% (Sigma, 30-264-3).

2.3. Chemicals for Liver Preconditioning

1. Retrorsine (Sigma, R0382).
2. Monocrotaline (Sigma, C2401).

2.4. Histochemical Staining for DPPIV

1. Gly-Pro-4-methoxy-β-naphtylamide (GPMNA) (Sigma, G9137), store at –20°C.
2. o-Dianisidine, tetrazotized (Fast Blue Salt BN) (Sigma, F3378), store at 4°C.
3. *N,N*-Dimethylformamide (Sigma, D4551).
4. Chloroform and acetone.
5. 0.1 M phosphate-buffered saline (PBS), pH 7.4.

2.5. Preparation of Monocrotaline Solution for Injection

Weigh suitable amounts of monocrotaline depending on body weight (dose 200 mg/kg) and dissolve in 0.9% saline in a small tube with the addition of 1 N HCl to lower pH around 2. After dissolution, adjust pH to 7 with 1 N NaOH. Pass the solution through a 22 μm filter.

2.6. Preparation of Retrorsine Solution for Injection

Dissolve retrorsine powder in PBS at 6 mg/ml, add up to two drops of 1 N HCl until complete dissolution, then adjust pH with 1 N NaOH to 7.4. Pass the solution through a 22 μm filter. Inject retrorsine in 500 μl per 100 g body weight.

2.7. Preparation of DPPIV Staining Solution

Dissolve 10 mg Fast Blue Salt BN in 10 ml 0.1× PBS in a small tube. Keep on ice. Dissolve 4 mg GPMNA in 0.5 ml *N,N*-dimethylformamide. Keep on ice. Combine solutions in a and b immediately before use. Freeze excess solutions in c at –70? for reuse over up to several months without repeated freeze-thawing.

2.8. Quantitative Analysis of Cell Proliferation

1. Spot RT digital camera (Diagnostic Instruments Inc., Sterling Heights, MI, USA) or equivalent.
2. ImageJ software (http://rsb.info.nih.gov/ij/).

3. Methods

3.1. Isolation of Rat Hepatocytes (21)

1. Set up a perfusion apparatus and maintain solutions at 37°C in a water bath (*see* **Note 1**).
2. Insert a 20 French intravascular catheter into the main portal vein and go beyond the tie but not beyond the bifurcation of the portal vein (*see* **Note 2**).
3. Start perfusion at 10–20 ml/min with 1× EGTA for 5 min (*see* **Note 3**).
4. Cut the abdominal aorta to drain perfusion fluid and continue perfusion with 1× Leffert's Buffer for 3 min and with collagenase-containing buffer for 10–20 min.
5. Excise the liver after completing perfusion, incise the capsule, and gently disperse cells.
6. Collect dissociated cell suspension and filter through 85 μm nylon mesh, followed by centrifugation of the cell suspension under 50×*g* for 5 min at 4°C at least twice to pellet viable cells and eliminate other cell types.
7. Resuspend cell pellet in RPMI 1640 medium, count cell numbers, and assess cell viability with trypan blue dye exclusion and maintain cells on ice for transplantation (*see* **Note 4**).

3.2. Transplantation of Isolated Hepatocytes in Rats

1. Anesthetize rat with anesthetic ether or other suitable medication, place in right decubitus position, and clean the abdominal wall with 70% ethanol and iodine.
2. Make 0.5–1 cm incision below the left subcostal abdominal wall with sharp scissors.
3. Identify the spleen and loosely tie a monofilament silk ligature around its lower pole.
4. Resuspend 10–20 million hepatocytes per milliliter of plain RPMI 1640 medium.
5. Inject 5–20 million cells through a 1 ml syringe with a 25 French needle into the lower pole of spleen over 10–20 s (*see* **Note 5**).
6. Withdraw the needle after injecting cells and tighten the silk ligature to prevent leakage of transplanted cells and bleeding.
7. Wipe blood with gauze and close the abdominal incision with 4-0 nylon sutures.
8. Return the animal to its cage, keep warm under heating lamp until recovery from anesthesia, and administer analgesia.

3.3. Preconditioning to Improve Cell Engraftment

1. Administer 200 mg/kg monocrotaline intraperitoneally or intravenously to DPPIV– rats 24 h before cell transplantation (*see* **Note 6**).

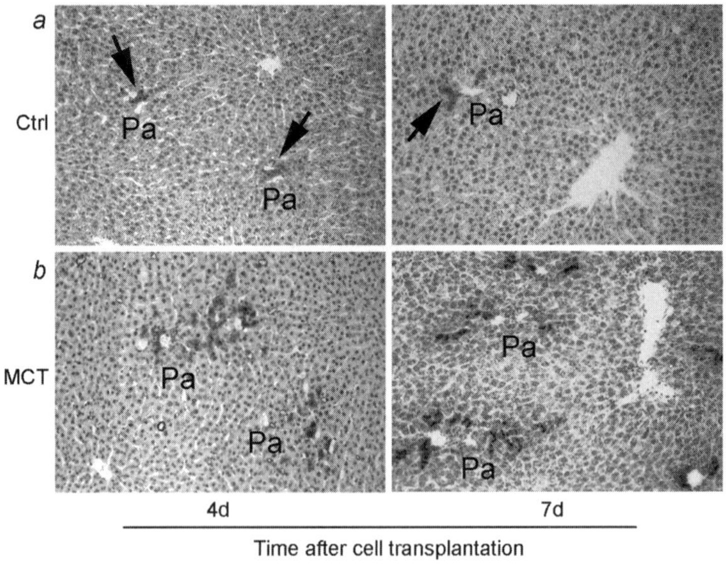

Fig. 10.1. **Liver preconditioning using monocrotaline (MCT) for improving cell engraftment in DPPIV– rats.** Transplanted F344 rat hepatocytes are shown in the recipient liver 4 and 7 days after cell transplantation. Panel **a** shows 1–3 transplanted hepatocytes with histochemically visualized DPPIV activity (red color, arrows) in periportal areas (Pa). By contrast, in MCT-treated rats (**b**) several-fold more transplanted cells are present. Original magnification, ×200; hematoxylin counterstain. Modified from Joseph B, et al. *(20)*. Reprinted with permission of Wiley-Liss, Inc., a subsidiary of John Wiley & Sons, Inc. (*see* Color Plate 4)

2. Analyze engraftment of transplanted cells 1, 2, 4, or 7 days or after longer intervals, e.g., 3 months, following cell transplantation (**Fig. 10.1**).

3.4. Preconditioning for Assessing the Proliferation of Transplanted Cells – Retrorsine Plus Partial Hepatectomy

1. Commence administration of 30 mg/kg retrorsine to DPPIV– rats when 6 weeks old weighing 70–100 g. Retrorsine is administered intraperitoneally in a volume of 500 μl per 100 g body weight. A second dose of retrorsine is given 2 weeks later.
2. Two-thirds partial hepatectomy is according to the standard Higgins and Anderson method *(22)*. Rats are anesthetized under ether, placed supine, and the abdominal wall is cleaned with 70% ethanol and iodine. A 2–3 cm midline laparotomy incision is made starting from the bottom of the xiphoid process. The falciform, and left triangular and lienorenal ligaments are divided. A nonabsorbable silk ligature is placed around the median lobe and left lateral lobe of the liver to enclose the inflow and outflow pedicles and tied securely underneath the liver. The liver parenchyma above the ligature is resected, blood is wiped, and the abdomen is closed with 4-0 nylon sutures. After injecting 0.5 ml warm saline intraperitoneally, the animal is returned to the cage and kept warm until recovery from anesthesia.

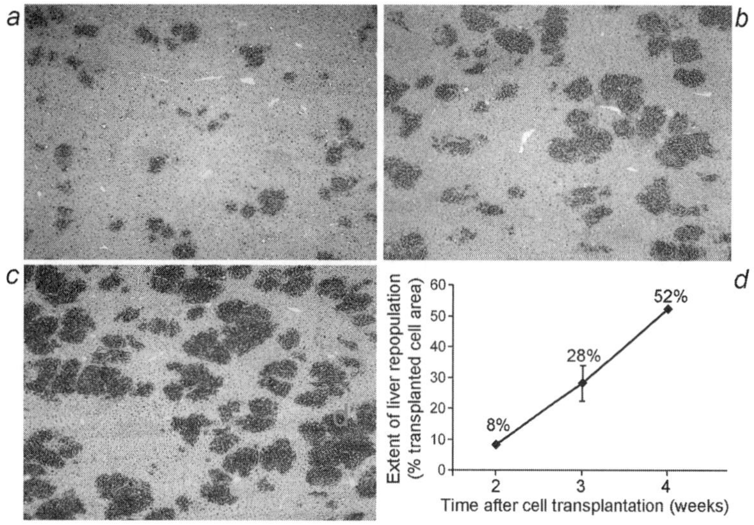

Fig. 10.2. **Analysis of the kinetics of liver repopulation in DPPIV– rats preconditioned with retrorsine and partial hepatectomy**. Foci of transplanted cells with DPPIV activity (red color) are seen 2 (**a**), 3 (**b**), and 4 weeks (**c**) after cell transplantation. Morphometric analysis of liver repopulation in panel **d** indicates linear increase in liver repopulation during this period. Original magnification, (**a–c**), ×40; hematoxylin counterstain. Modified from Wu Y-M et al. *(18)*. Reprinted with permission of Wiley-Liss, Inc., a subsidiary of John Wiley & Sons, Inc. (*see* Color Plate 5)

3. To assess liver repopulation, 5 million F344 hepatocytes are transplanted intrasplenically immediately after or 4–7 days after partial hepatectomy followed by timed analysis of livers.
4. To demonstrate the kinetics of liver repopulation, hepatocyte recipients are analyzed 2, 3, and 4 weeks following cell transplantation (**Fig. 10.2**).

3.5. Identification of Transplanted Hepatocytes by DPPIV Histochemical Staining

1. Tissues are sampled from multiple lobes per animal and frozen in methylbutane cooled to –70°C on dry ice. Cryosections of 5–6 μm thickness are prepared.
2. The sections are air-dried for at least 30 min and fixed in chloroform-acetone (1:1, v/v) at 4°C for 10 min.
3. The sections are covered with 50–100 μl staining solution and incubated for 30–45 min at room temperature in humidified chambers.
4. The staining solution is removed and sections are washed with clean water before counterstaining with aqueous 0.5% methylgreen or toluidine blue for 10–30 s.
5. Stained slides can be stored at 4°C after air-drying without mounting medium.
6. For microscopic examination and microphotography, stained sections are mounted in aqueous medium, e.g., glycerol.

3.6. Quantification of Engrafted Cells

1. Analyze multiple cryosections per liver lobe after DPPIV staining.
2. Place one reinforcement ring (3.7 mm in diameter) on either the coverslip or the back of the slide for the hole to not extend beyond the tissue section.
3. Count transplanted cells identified by DPPIV staining under ×400. Analyze 100 consecutive liver lobules. Assess the fraction of portal vein radicles containing transplanted cells.
4. The number of transplanted cells can be depicted as transplanted cells per liver lobule, as well as transplanted cells per unit liver area (mm^2) or volume (mm^3).

3.7. Measuring the Extent of Liver Repopulation by Transplanted Cells

1. Analyze multiple sections per liver lobe following DPPIV staining and obtain microphotographs from consecutive adjacent areas in sections under ×40 magnification.
2. To estimate the fraction of liver repopulated by transplanted cells, use ImageJ, or equivalent software, according to instructions.

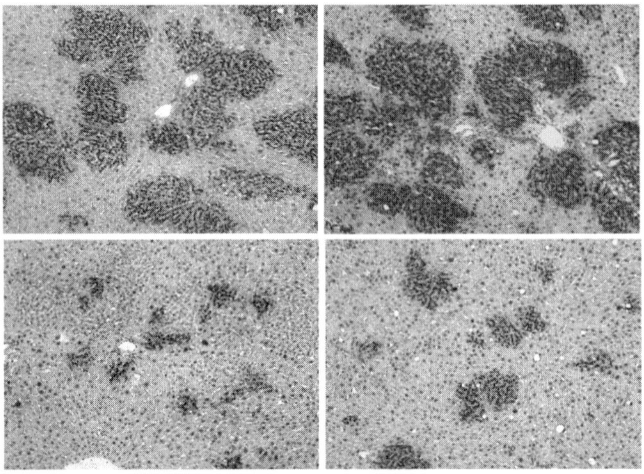

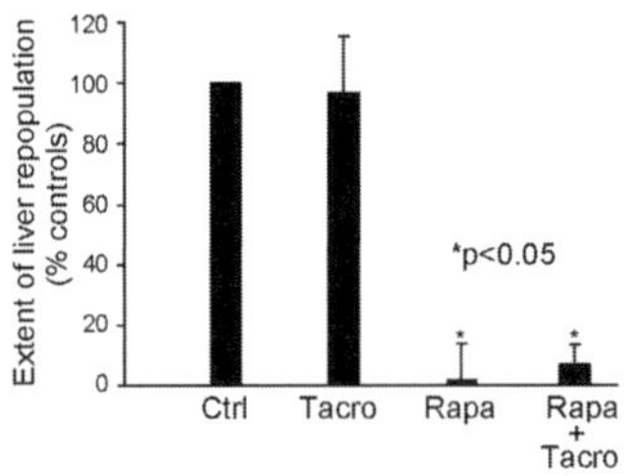

f

Condition (n = 3-5 each)	Size of transplanted cell foci ($x10^{-3}$ mm^3)	P values vs. controls
Controls	176±12	-
Tacro	169±31	0.99
Rapa	61±8	0.002
Rapa + Tacro	62±7	0.003

Fig. 10.3. **Effect of immunosuppressive drugs, Rapamycin (Rapa) and Tacrolimus (Tacro), on liver repopulation in DPPIV– rats preconditioned with retrorsine and partial hepatectomy.** Animals were treated with drugs subsequent to the completion of cell engraftment. Rapa- but not Tacro-suppressed transplanted cell proliferation as shown by DPPIV histochemistry and morphometric analysis of either the extent of liver repopulation (**e**) or individual transplanted cell foci (**f**). Original magnification (**a–d**), ×100; hematoxylin counterstain. Modified from Wu Y-M et al. *(18)*. Reprinted with permission of Wiley-Liss, Inc., a subsidiary of John Wiley & Sons, Inc. (see Color Plate 6)

3. To estimate the size of transplanted cell foci in microphotographs, use Neubaur chamber (1 mm^2 per square) or equivalent for calibration.

Using this preconditioning protocol for repopulation of the liver, the behavior of transplanted cells can be conveniently demonstrated. As an illustration, studies to analyze the effect of the immunosuppressive drug, Rapamycin, on transplanted hepatocytes were informative (**Fig. 10.3**). These studies *(18)* demonstrated that Rapamycin suppressed proliferation in transplanted cells, resulting in the arrest of liver repopulation. This analysis was helpful in establishing which immunosuppressive drugs will be most suitable for clinical liver cell therapy protocols.

4. Notes

1. Filter all solutions after adjusting pH to 7.4. Prepare collagenase solution immediately before perfusion.
2. Advancement of the IV catheter beyond the bifurcation of the portal vein will produce variable liver perfusion.
3. After portal cannulation, perfusion should be started immediately to avoid thrombotic occlusion of distal portal vein radicles, which impairs perfusion and tissue digestion. Low-dose heparin intravenously before portal cannulation may be helpful for the beginner.
4. The viability of isolated hepatocytes will affect cell engraftment and proliferation. Hepatocytes with viability less than 80% should not be used for transplantation.
5. Use of large volumes (e.g., in excess of 2 ml) may produce splenic rupture or hemorrhage.
6. Proliferation of transplanted cells after preconditioning with retrorsine and partial hepatectomy is affected by the gender. Male rats are more responsive to both retrorsine and monocrotaline.

References

1. Ponder, K. P., Gupta, S., Leland, F., et al. (1991) Mouse hepatocytes migrate to liver parenchyma and function indefinitely after intrasplenic transplantation. *Proc Natl Acad Sci USA* 88, 1217.
2. Gupta, S., Aragona, E., Vemura, R. P., et al. (1991) Permanent engraftment and function of hepatocytes delivered to the liver: implications for gene therapy and liver repopulation. *Hepatology* 14, 144.
3. Harmeet, M., Pallavi, A., Sanjeev, S., et al. (2002) Cyclophosphamide disrupts hepatic sinusoidal endothelium and improved transplanted cell engraftment in rat liver. *Hepatology* 36, 112–121.
4. Kim, K. S., Joseph, B., Inada, M., et al. (2005) Regulation of hepatocyte engraftment and proliferation after cytotoxic drug-induced perturbation of the rat liver. *Transplantation* 80, 653–659.

5. Sanjeev, S., Pankaj, R., Yoshiya, I., et al. (2002) Hepatic sinusoidal vasodilators improve transplanted cell engraftment and ameliorate microcirculatory perturbations in the liver. *Hepatology* 35, 1320–1328.
6. Kumaran, V., Joseph, B., Benten, D., et al. (2005) Integrin and extracellular matrix interactions regulate engraftment of transplanted Hepatocytes in the rat liver. *Gastroenterology* 129, 1643–1653.
7. Brigid, J., Harmeet, M., Kuldeep, K. B., et al. (2002) Kupffer cells participate in early clearance of syngeneic hepatocytes transplanted in the rat liver. *Gastroenterology* 123, 1677–1685.
8. Rajvanshi, P., Kerr, A., Bhargava, K. K., et al. (1996) Efficacy and safety of repeated hepatocyte transplantation for significant liver repopulation in rodents. *Gastroenterology* 111, 1092–1102.
9. Yuan, R. H., Ogawa, A., Ogawa, E., et al. (2003) p27Kip1 inactivation provides a proliferative advantage to transplanted hepatocytes in DPP?/Rag2 double knockout mice after repeated host liver injury. *Cell Transpl* 12, 907–919.
10. Rhim, J. A., Sandgren, E. P., Degen, J. L., et al. (1994) Replacement of diseased mouse liver by hepatic cell transplantation. *Science* 263, 1149–1152.
11. Alexandre, M., Jacques, E. G., Claudia, M., et al. (1998) Selective repopulation of normal mouse liver by Fas/CD95-resistant hepatocytes. *Nat Med* 4, 1185–1188.
12. Overturf, K., Al-Dhalimy, M., Ou, C. N., et al. (1997) Serial transplantation reveals the stem-cell-like regenerative potential of adult mouse hepatocytes. *Am J Pathol* 151, 1273–1280.
13. Harmeet, M., Giridhar, R. G., Adil, N. I., et al. (2002) Cell transplantation after oxidative hepatic preconditioning with radiation and ischemia-reperfusion leads to extensive liver repopulation. *Proc Nat Acad Sci USA* 99, 13114–13119.
14. Guha, C., Yamanouchi, K., Jiang, J., et al. (2005) Feasibility of hepatocytes transplantation-based therapies for primary hyperoxalurias. *Am J Nephrol* 25, 161–170.
15. Ezio, L., Ran, O., Deb, K. M., et al. (1998) Long-term, near-total liver replacement by transplantation of isolated hepatocytes in rats treated with retrorsine. *Am J Pathol* 153, 319–329.
16. Oren, R., Dabeva, M. D., Karnezis, A. N., et al. (1999) Role of thyroid hormone in stimulating liver repopulation in the rat by transplanted Hepatocytes. *Hepatology* 30, 903–913.
17. Guo, D., Fu, T., Nelson, J. A., et al. (2007) Liver repopulation after cell transplantation in mice treated with retrorsine and carbon tetrachloride. *Transplantation* 73, 1818–1824.
18. Wu, Y. M., Joseph, B., Gupta, S. (2006) Immunosuppression using the mTOR inhibition mechanism affects replacement of the rat liver with transplanted cells. *Hepatology* 44, 410–419.
19. Witek, R. P., Fisher, S. H., Petersen, B. E. (2005) Monocrotaline, an alternative to retrorsine-based hepatocytes transplantation in rodents. *Cell Transpl* 14, 41–47.
20. Joseph, B., Kumaran, V., Berishvili, E., et al. (2006) Monocrotaline promotes transplanted cell engraftment and advances liver repopulation in rats via liver conditioning. *Hepatology* 44, 1411–1420.
21. Neufeld, D. S. (1997) Isolation of rat liver hepatocytes. *Methods Mol Biol* 75, 145–151.
22. Higgins, G. M., Anderson, R. M. (1931) Experimental pathology of liver resection. *Arch Pathol* 12, 186–197.

Chapter 11

Ex Vivo Gene Transfer into Hepatocytes

Xia Wang, Prashant Mani, Debi P. Sarkar, Namita Roy-Chowdhury and Jayanta Roy-Chowdhury

Abstract

Ex vivo gene transfer into hepatocytes could serve several purposes in the context of gene therapy or cell transplantation: *(1)* isolated hepatocytes can be transduced in culture with therapeutic genes and then transplanted into the recipient; *(2)* marker genes can be introduced for subsequent identification of transplanted cells and their progeny; *(3)* gene transfer can be used for conditional immortalization of hepatocytes for expansion in culture; *(4)* immunomodulatory genes can be transferred into hepatocytes to prevent allograft rejection. Gene transfer into cultured hepatocytes can be achieved using DNA that is not incorporated into recombinant viruses. In such systems, transgene integration into the host cell genome can be enhanced using transposon systems, such as "sleeping beauty." In addition to using the conventional reagents, such as cationic liposomes, DNA transfer into hepatocytes can be achieved by Nucleofection® or special hepatocyte-targeted carriers such as proteoliposomes containing galactose-terminated glycoproteins (e.g. the F protein of the Sendai virus). Alternatively, genes can be transferred using recombinant viruses, such as adenoviral vectors that are episomal or retroviral vectors (including lentiviruses) that permit integration of the transgene into the host genome. Gene transfer using lentiviral vectors has been achieved in both attached and suspended hepatocytes. Transduction efficiency of lentiviral vectors can be enhanced using magnetic nanoparticles (Magnetofection®).

Key words: Gene transfer, ex vivo, sleeping beauty, nucleofection, F-virosome, lentiviral vectors, magnetofection.

1. Introduction

Transferring genes into isolated hepatocytes could enhance the scope of hepatocyte transplantation. Some examples of the potential uses of ex vivo gene transfer are discussed below to illustrate that the choice of methods to transfer the transgene depends on the ultimate goal of the procedure.

Anil Dhawan, Robin D.Hughes (eds.), *Hepatocyte Transplantation, vol. 481*

DOI 10.1007/978-1-59745-201-4_11 Springerprotocols.com

1.1. Objectives of Ex Vivo Gene Transfer into Hepatocytes

1.1.1. Ex Vivo Gene Therapy

This procedure consists of isolating hepatocytes from a patient or a mutant animal carrying a liver-based inherited disease, transducing the cells in culture with a therapeutic gene and then transplanting the phenotypically corrected cells back into the donor. Since the hepatocytes are autologous, this approach circumvents the need for immunosuppression of the host. Long-term efficacy of this strategy requires integration of the transgene into the host genome.

Hepatocytes from a resected liver segment from low-density lipoprotein (LDL) receptor-deficient rabbits (Watanabe heritable hyperlipidemic rabbit) have been transplanted after ex vivo transduction with the low-density lipoprotein receptor (*LDLR*) gene using recombinant Moloney's murine leukemia virus (MuLV) vectors *(1)*. This study and the subsequent clinical trial in human subjects *(2)* with familial hypercholesterolemia had only a minor metabolic effect, which was not sufficient for clinical benefit. Several technical issues limited the success of the procedure. (i) Because cultured primary hepatocytes do not proliferate significantly and have a limited life span in culture, it was not possible to select the transduced cells prior to transplantation. Therefore, the success of the procedure was dependent primarily on the efficiency of transduction. Oncoretroviruses, such as MuLV, require cell division for integration into the chromosome. Despite the use of growth factors in the media, there was only a minor degree of mitosis of hepatocytes in culture. Thus, the efficiency of transduction was limited. (ii) The number of hepatocytes that can be safely transplanted in a single procedure is limited. As no preparative maneuver had been employed to promote preferential proliferation of the transplanted cells in the host liver, the total number of engrafted phenotypically corrected hepatocytes was quite small. Nonetheless, these studies demonstrated that the procedure can be performed safely and delineated the problems involved in this approach, which has stimulated further research, addressing each hurdle as described below.

Vectors, such as those based on immunoretroviruses (lentiviruses) and plasmids that are transposition competent, exhibit a high efficiency of integration in non-dividing cells. Substitution of the oncoretroviral vectors with these vectors could provide a high level of gene transfer into primary hepatocytes, enabling successful ex vivo gene therapy, without the need for prior selection of the transduced cells. New development in the area of hepatic repopulation with transplanted hepatocytes, such as those based on controlled irradiation of the host liver and the use of hepatocyte mitostimulatory factors, can enable preferential proliferation of the engrafted cells over host hepatocytes, leading to progressive repopulation of the liver.

1.1.2. Marking Hepatocytes for Imaging

Currently, the assessment of survival of the engrafted hepatocytes in the host liver requires needle biopsies or surgical biopsies of the liver, which is difficult and risky to repeat in clinical settings and can be misleading because of the inhomogeneous distribution of the engrafted cells. A major obstacle to improving the techniques for hepatocyte transplantation is the lack of a non-invasive method for serial assessment of the survival and distribution of the transplanted hepatocytes. In small animals, optical imaging employing hepatocytes expressing firefly luciferase or green fluorescent protein can be used for localization of engrafted cells (*3*, *4*). But optical methods do not offer the degree of penetration that would be needed in larger animals or humans, because of the thickness of the abdominal wall. Positron emission tomography (PET) and single photon emission computed tomography (SPECT) are sensitive enough for human studies *(5)*. The above methods require the expression of non-mammalian gene products, which may be immunogenic or may have other toxic effects. Recently, it has been possible to determine the distribution of the engrafted donor hepatocytes expressing creatinine kinase (CK) and to quantify the extent of hepatic repopulation using magnetic resonance spectrometric imaging (MRSI) *(6)*. CK is not expressed constitutively in hepatocytes. CK-mediated phosphorylation of creatinine (Cr) in the donor cells produces phosphocreatine (PCr) that is absent in normal liver, thereby generating a specific ^{31}P NMR spectrum in vivo.

For the assessment of initial engraftment, it may be sufficient to employ vectors that do not lead to integration of the marker gene. However, long-term assessment, especially after repeated mitosis of the engrafted cells, transgene integration into donor cell chromosomes is necessary.

1.1.3. Conditional Immortalization of Hepatocytes

The shortage of donor organs has prompted investigators to design strategies for conditional immortalization of hepatocytes. The strategies involve the use of immortalizing gene products that may be degraded rapidly at physiological temperatures (e.g. thermolabile Simian Virus 40 T-antigen) or transgenes that can be removed before or after engraftment (e.g. T-antigen flanked by P-lox sequences). Conditionally immortalized hepatocytes have been used successfully in rodents to provide metabolic support during acute *(7)* or chronic liver failure *(8)*. Conditional immortalization requires integration of the transgene into the hepatocyte genome.

1.1.4. Prevention of Immune Rejection of the Transplanted Hepatocytes

Available methods for preventing allograft rejection involve the suppression of the host immune system by the use of immunosuppressive agents, which are associated with many untoward side effects. Expression within the donor cells of non-secreted gene products that could prevent allorejection without modulating the host immune system could represent a major advance toward the clinical application of hepatocyte transplantation. Recently, expression of certain viral gene products within donor hepatocytes has been shown to prevent their allograft rejection by protecting the engrafted cells from the effector limb of the host alloimmune response *(9)*. This approach probably requires the integration of the transgene into the host genome.

1.2. Approaches to Ex Vivo Gene Transfer into Hepatocytes

1.2.1. Constructing Plasmids Expressing the Gene of Interest

An important consideration in designing plasmids for *ex vivo* gene transfer is promoter selection. Several viral promoters, including the cytomegalovirus immediate early promoter (CMV-IE), have been reported to be silenced over time, particularly when the transcription unit is integrated into the hepatocyte genome. Therefore, selection of a ubiquitous vertebrate promoter (e.g. the phosphoglycerate kinase (PGK), chicken β-actin or the eukaryotic initiation factor 1A promoter) or a hepatocyte-specific promoter (albumin, α-fetoprotein or α_1-antitrypsin promoter) may be preferred. Hepatocyte-specific promoters provide the advantage of restricting the gene expression to hepatocytes, which may reduce the immune response against the expressed protein. However, some hepatocyte-specific promoters, such as the albumin promoter, are downregulated during an inflammatory response, which occurs in the liver immediately after hepatocyte transplantation. Generally, a “strong” promoter is chosen to maximize the transgene expression, although there are exceptions to this. In the case of transposition-enabled plasmids expressing the *sleeping beauty* transposase (see below), expression of the transposes from a “weak” promoter is desirable, because excessive amount of the transposase inhibits transposition.

Although plasmid transfection into hepatocytes generally results in the transient expression of transgenes, in recent years, transposition-enabled plasmids have been designed to increase markedly the frequency of integration of transgenes into the host genome *(10, 11)*. In this system, the transcription unit of interest is flanked by inverted and direct repeats that are binding sites for the *sleeping beauty* transposase. The transposase may be expressed from a second transcription unit on the same plasmid (*cis*) or from a different plasmid that is cotransfected (*trans-*) with the plasmid containing the gene of interest. As discussed above, it is important to keep the expression of the transposase at a low level to achieve high levels of integration of the gene of interest *(12)*.

1.2.2. Gene Transfer Using DNA That Is Not Incorporated into Recombinant Viruses

Plasmids or other forms of DNA, unincorporated in viral vectors, can be transfected into primary hepatocytes. Generally, such transfections result in transient expression of the gene and integration into the cellular genome occurs infrequently. When selection of the stably transfected cells is possible, such as in the case of transfection with immortalizing genes, simple transfection can be effective. Hepatocytes are more susceptible to injury by many standard transfection vehicles than are other cell types. Calcium phosphate coprecipitation, diethylaminoethyl-dextran and conventional electroporation methods of transfection result in an unacceptable degree of cell death. However, new transfection methods (such as Nucleofection®, Amaxa, Gaithersburg, MD, USA) that combine a chemical and an electroporation approach have been successful. DNA can be transfected into hepatocytes using liposomes as carriers. In an effort to provide hepatocyte specificity of transfection, so that other contaminating cells in the preparation are not transfected inadvertently, hepatocyte-specific ligands have been used as the transfection vehicle. Galactose-terminated asialoglycoproteins, such as asialofetuin or asialo-orosomucoid, have been conjugated with a polycation to serve as a vehicle to transfer the DNA by endocytosis via the hepatocyte-specific asialoglycoprotein receptor (ASGR). However, molecules transferred by this pathway are naturally targeted to the lysosome, where they are degraded. This reduces the transfection efficiency. As an ingenious solution to this problem, investigators have used the F-protein of the Sendai virus as a hepatocyte-specific ligand. The F-protein has a high affinity for ASGR, but its fusogenic activity leads to delivery of the cargo to the cytosol, rather than to the endosomes *(13)*. Some protocols for transfection of DNA without incorporation into recombinant viruses are described below. This is not a comprehensive list, but covers methods that we have found to be useful in our laboratory.

1.2.3. Gene Transfer Using Recombinant Viruses

Recombinant viruses can greatly enhance gene transfer efficiency. Where transient gene transfer is needed, recombinant adenoviruses offer the most effective means of ex vivo gene transfer into hepatocytes. However, when integration into the host chromosome is required for persistence of transgene expression into the progeny of the transduced cells, recombinant retroviruses, including lentiviral vectors, may be employed.

2. Materials

2.1. Hepatocyte Culture

1. RPMI 1640 (Invitrogen, Carlsbad, CA) supplemented with 10% fetal bovine serum (FBS, Sigma, St. Louis, MO, USA), 100 μg/ml streptomycin, 100 U/ml penicillin (Invitrogen, Carlsbad, CA, USA), 2 mM L-glutamine (Fisher, Pittsburgh, PA, USA)
2. Dulbecco's Modified Eagle's Medium (DMEM) (Fisher) supplemented with 10% FBS, 100 μg/ml streptomycin, 100 U/ml penicillin.
3. Iscove's DMEM (IMDM, Invitrogen) contains 10% FBS, 100 μg/ml streptomycin and 100 U/ml penicillin.
4. OptiMEM (Invitrogen).
5. Dexamethasone is dissolved in 100% ethanol at 25 μmol/l and stored in –20°C. Final Dexamethasone concentration should be 25 nmol/L.
6. Bovine insulin (Sigma). Working concentration is 5 μg/ml.
7. Hexadimethrine bromide (Polybrene, Fisher) is dissolved at 8 mg/ml in ddH_2O and stored in –20°C. Final working concentration of polybrene is 8 μg/ml.
8. Collagen (Vitrogen 100, Cohesion Technologies, Palo Alto, CA, USA). Collagen is diluted with 0.012 N HCl (*see* **Note 1**).

2.2. Nucleofection by Amaxa

1. Mouse hepatocyte Nucleofector® solution (Amaxa, Gaithersburg, MD, USA) (*see* **Note 2**).
2. Amaxa®-certified cuvette (Amaxa).

2.3. F-Virosome

1. Inactivated Hemaglutinating virus of Japan (Sendai virus-Z strain, HVJ) (Charles River, North Franklin, CT, USA).
2. 0.02 M Tris-buffered saline (pH 8.3). Store at room temperature.
3. Dithiothreitol (DTT, Sigma) is dissolved in 0.02 M Tris-buffered saline (pH 8.3) at 30 mM, always make fresh (*see* **Note 3**), the final concentration should be 3 mM.
4. Dialysis bag (cutoff MW 12,000–14,000, VWR, Batavia, IL, USA).
5. 10 mM Tris-buffered saline, pH 7.4.
6. 10% Triton X-100, keeping the final percentage of triton between 2 and 5%.
7. SM-2 Biobeads (Bio-Rad, Hercules, CA, USA).
8. 26-Gauge needle (Fisher).
9. Histidine containing lipid and control lipid (Provided by Dr. A. Chaudhuri, Indian Institute of Chemical Technology, Hyderabad) is dissolved in solvent (methanol/chloroform, 2:1).

2.4. Lentiviral Vectors

1. 2× BBS (150 mM NaCl; 50 mM Bes; and 1.5 mM Na_2HPO_4 pH 6.95).
2. 0.22 µm filter (Fisher).
3. 2.5 M $CaCl_2$ is filtered with a 0.22-µm filter, stored in −20°C.

2.5. Tissue Staining for β-Galactosidase

1. 100% Ethanol.
2. 5-Bromo-4-chloro-3-indolyl-b-D-galactopyranoside (X-Gal, Fisher) is dissolved at 40 mg/ml in dimethyl sulfoxide (DMSO, Sigma) and stored in −20°C. Complete X-Gal staining solution contains 1 mg/ml X-Gal, 35 mM potassium ferricyanide, 35 mM potassium ferrocyanide, 2 mM $MgCl_2$.

2.6. Immunostaining

1. Microscope coverslips (Fisher).
2. Fixation solution: acetone/methanol (1/4, v/v).
3. 70% Ethanol in phosphate-buffered saline (PBS).
4. Antibody dilution buffer: 1% bovine serum albumin (BSA) and 0.5% Tween-20 in PBS.
5. Primary antibody WP1 is diluted at 1:10 antibody dilution buffer.
6. Secondary antibody: anti-mouse IgG conjugated to alkaline phosphatase (AKP) (Sigma) is diluted at 1:200 antibody dilution buffer.
7. Alkaline Phosphatase substrate kit III (Vector Laboratories, Burlingame, CA, USA). One drop of A, B and C each are added to 2.5 ml pH 8.2 Tris-HCl.

2.7. Magnetofection Enhancement of Viral Vector-Mediated Gene Transfer

ViroMag® and ViroMag R/L® (Oz Biosciences, Marseille, France). (*see* **Note 4**).

2.8. Adenoviral Vectors

1. 10 mM Tris at pH 8.0.
2. 1,1,2-Trichloro-1,2,2-trifluoroethane (Freon, Fisher). (*see* **Note 5**).
3. CsCl gradient: 67 g CsCl is dissolved in 100 ml 10 mM Tris (pH 8.0) as 1.4p 1.2p CsCl can be prepared by adding the same volume 10 mM Tris (pH 8.0) to 1.4p CsCl.
4. 1% sodium dodecyl sulfate (SDS).
5. Fixation solution: 100% ice-cold methanol.
6. Mouse anti-hexon antibody (BD, Franklin Lakes, NJ, USA).
7. Rat anti-mouse antibody conjugated to horseradish peroxidase (HRP) (BD).
8. Diaminobenzidine substrate (DAB, BD).

3. Methods

3.1. Primary Hepatocytes Preparation for Gene Transfer

Hepatocytes are isolated by in situ perfusion of the liver by minor modifications *(14)* of the method originally described by Berry and Friend *(15)*. The method for rat liver perfusion is described here, but the method can be adapted for both smaller and larger animal livers. Modification of this method for perfusion of resected liver segments has been described elsewhere *(16)*. 1.5×10^6 primary hepatocytes are plated onto 100 mm pre-coated plates. These cells are then cultured in DMEM with 10% fetal calf serum (FBS), 100 μg/ml streptomycin, 100 U/ml penicillin, 2 mM L-glutamine, 1 mM sodium pyruvate, 25 nM dexamethasone and 5 μg/ml bovine insulin.

3.2. Protocol for Gene Transfer into Hepatocytes Using Conventional Liposomes

3.2.1. Preparation of Cationic Liposome–DNA Complex (17–20)

There are several effective liposome preparations that are available commercially. A typical example is given below.

1. The optimized concentration of the gene of interest cDNA (1–20 μg) is incubated at room temperature (22°C) with the liposome (1–50 μg) for 20 min by gentle mixing. The liposome consists of a 3:1 formulation of 2,3-dioleyloxy-N-[2 (sperminecaboxamido)ethyl]-*N*-*N*-dimethyl-1-propanaminiumtrifluoroacetate (DOPSA) and dioleoylphosphatidyl ethanolamine (DOPE).
2. After incubation, a final concentration of 16 μg DNA and 40 μg liposome is obtained by dilution with OptiMEM (Invitrogen).

3.2.2. Transfection of Hepatocytes

1. Primary hepatocytes are rinsed in Dulbecco's phosphate-buffered saline (DPBS) and incubated with the liposome–DNA complex for 2 h at 37°C.
2. After this initial incubation period, complete DMEM is added to the cells, which are incubated for an additional 24 h.
3. Twenty hours after transfection, the cells are washed with DPBS and released from the plate by incubation with 0.25% trypsin for 5 min and sedimented by centrifugation ($50\times g$ for 5 min) at 4°C. The cell pellet is reconstituted in 500 μl of PBS for transplantation.

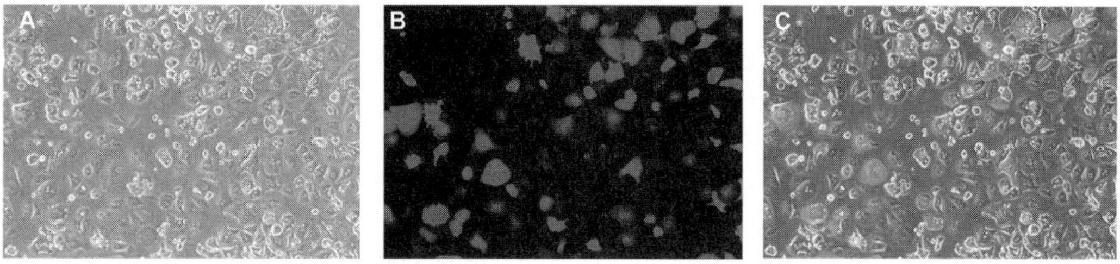

Fig. 11.1. **Transfection by Amaxa Nucleofection:** Expression of GFP in primary mouse hepatocytes (isolated from C57BL/6 mice) nucleofected using an Amaxa mouse hepatocyte Nucleofector kit with a plasmid encoding maxGFP. Twenty-four hours after nucleofection, cells were analyzed by bright field (**A**) and fluorescence microscopy (**B**). The merged image is shown in panel (**C**). (*see* Color Plate 7)

3.3. Protocol for Nucleofection (Amaxa, Gaithersburg)

This method combines the principles of chemical transfection and electroporation. Methods have been optimized by the manufacturer for various types of cells *(21)*, including hepatocytes (**Fig. 11.1**). The transfection buffers (Nucleofector solutions) are proprietary and their compositions are not published. The following protocol that has been developed by Amaxa has been validated in our laboratory.

Protocol for transfecting primary hepatocytes from rats or mice:

1. Warm the supplemented mouse hepatocyte Nucleofector solution to room temperature.
2. Pre-equilibrate RPMI 1640 (Invitrogen) with 10% FBS, 100 μg/ml streptomycin, 100 U/ml penicillin in a humidified 37°C incubator containing air with 5% CO_2.
3. Isolated primary mouse or rat hepatocytes are sedimented at $50\times g$ for 5 min and then gently resuspended with mouse hepatocyte Nucleofector solution at 7×10^5 cells/100 μl at room temperature. The cell suspension is not kept in the Nucleofector solution for more than 15 min before Nucleofection as this reduces cell viability.
4. Mix 100 μl of cell suspension with 2–6 μg plasmid DNA containing the gene of interest.
5. Transfer the mixture into an Amaxa-certified cuvette. The sample should cover the bottom of the cuvette and there should be no air bubbles. The cuvette is capped.
6. Select the approriate Nucleofector program, T-28 or T-028 (*see* Nucleofector I or Nucleofector II Manual for details). Insert the cuvette into the cuvette holder (for Nucleofector I: rotate carousel to final position) and press the "X" button to start the program.
7. After completion of the program the sample in the cuvette is incubated for 15 min at room temperature and 500 μl of the pre-equilibrated culture medium is added to

the cuvette. The cells are then gently transferred to six-well plates.

8. Incubate cells in a humidified 37°C /5% CO_2 incubator.
9. After 4 h, replace the medium with fresh complete DMEM (hepatocytes should be attached to the plate at this point of time).
10. Expression of the transgene is evaluated 24–48 h after Nucleofection. (*see* **Note 6**): Transgene expression can be detected within 6 h after Nucleofection. The rapid expression of the transgene is presumably due to the more efficient transfer of the DNA to the nucleus than with other methods of gene transfer.

3.4. Transfection Using Liposomes Containing Components of the Sendai Virus

The hemagglutinating virus of Japan (HVJ, also called Sendai virus), an enveloped paramyxovirus, has a long history of being utilized for its cell fusion properties. Liposomes derived by detergent solubilization of the virus contain two major glycoproteins, hemagglutinin neuraminidase (HN) and fusion factor (F). Liposomes generated from detergent-solubilized Sendai viruses lack the viral genome but contain HN and F. Such liposomes can be used to entrap exogenous DNA for transfer into a wide variety of cell types. Binding of HN proteins to the cell membrane promotes fusion of the complex liposome to the cell membrane, which is mediated by the F-protein. Liposomes containing both HN and F proteins lack host cell specificity and exhibit some cell toxicity. To design DNA transfection vehicles that are targeted to hepatocytes, biochemical methods have been developed to eliminate the HN protein from the complex. Briefly, treatment of the virus with a strong reducing agent (e.g. dithiothretol), followed by removal of the reducing agent leads to regeneration of F-protein, but to irreversible denaturation and insolubilization of the HN protein. Subsequent detergent solubilization of the virus and addition of the DNA of interest, followed by removal of the detergent, leads to the formation of liposomes composed of virus-derived lipids, the F-protein and the entrapped DNA of interest. The galactose-terminated carbohydrate moieties with fucose side chains confer the F-protein with high affinity and specificity for hepatocyte-specific ASGRs. Thus, the F-protein-containing liposomes (termed F-virosomes) are hepatocyte-specific. However, the absence of HN in F-virosomes reduces the gene transfer efficiency. Although the mechanism by which HN enhances the fusion activity of F-virosomes is not understood completely, histidine residues of HN are known to be important for this function. Incorporation of histidinylated lipids in the F-virosome–DNA complex is thought to confer several benefits, including compaction of the DNA, enhancement of the packaging capacity of individual F-virosomes, augmenting the fusogenic

Color Plates

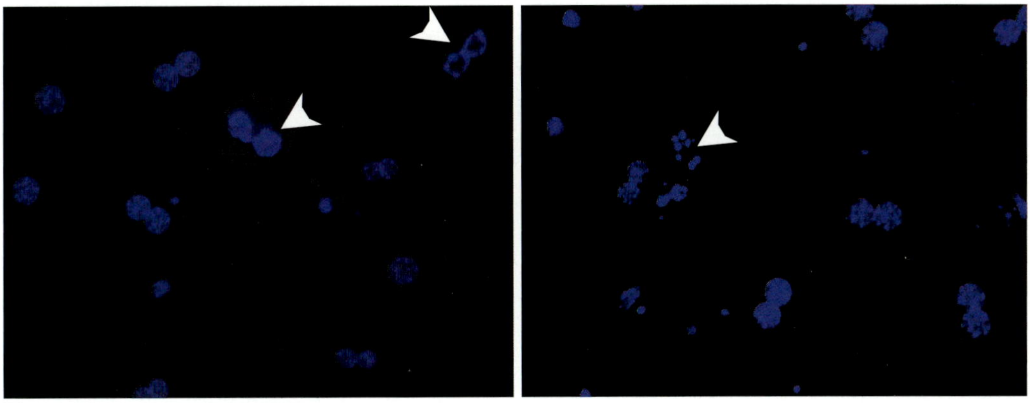

Color Plate 1. Apoptotic nuclei and bodies observed in mouse primary hepatocyte cultures after staurosporine treatment (white arrows). Freshly isolated mouse hepatocytes were plated for 24 h on a collagen type I-coated coverslips in well plates and treated for 4 h with 1 μM staurosporine. Cells were thereafter fixed with 4% of formaldehyde for 20 min at room temperature, stained with DAPI for 30 min and analyzed using a fluorescence microscopy. (*see* discussion on p. 63)

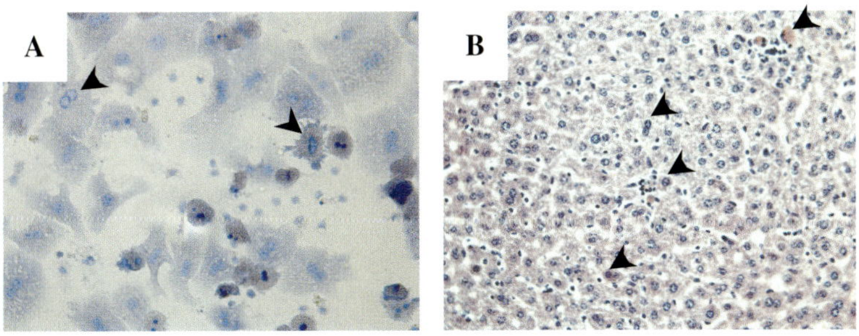

Color Plate 2. Condensation of chromatin at the periphery of the nucleus in apoptotic mouse hepatocytes (black arrows). (A) Primary mouse hepatocytes were plated for 24 h in a coated collagen type I well plates and treated for 4 h with 1 μM staurosporine. Cells were thereafter fixed with 4% formaldehyde for 20 min at room temperature and stained with HE for 10 min. (B) slice of mouse liver prefixed with formaldehyde, paraffin-embedded and HE-stained. (*see* discussion on p. 65)

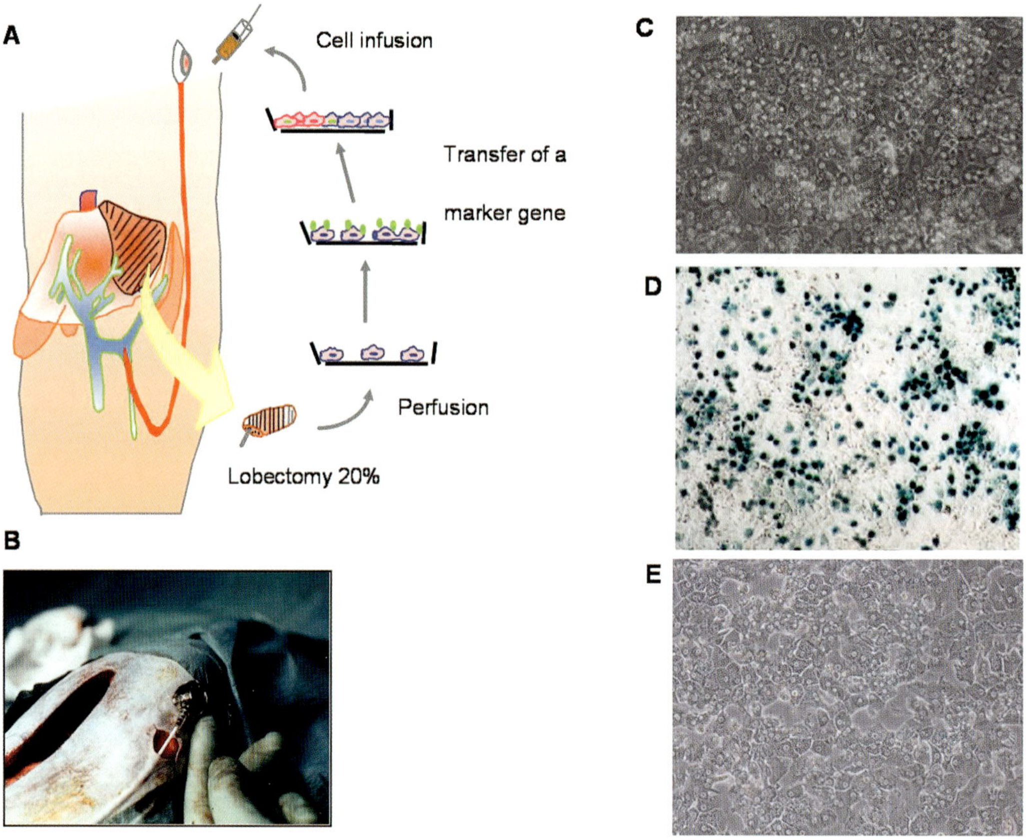

Color Plate 3. Transplantation of autologous hepatocytes into *Macaca mulatta* after retroviral-mediated gene marking. (**A**) Protocol for simian hepatocyte isolation, retroviral transduction and transplantation. Hepatocyte transduction with HIV-1-derived lentivirus vectors avoids the culture steps. They are transduced in suspension and transplanted. (**B**) Hepatocytes are transplanted via the infusion chamber. (**C**) Freshly isolated simian hepatocytes at confluency after 3 days of culture. (**D**) Transduced hepatocytes in culture expressing the β-galactosidase. (**E**) Thawed hepatocytes after 3 days of culture. (*see* discussion on p. 90)

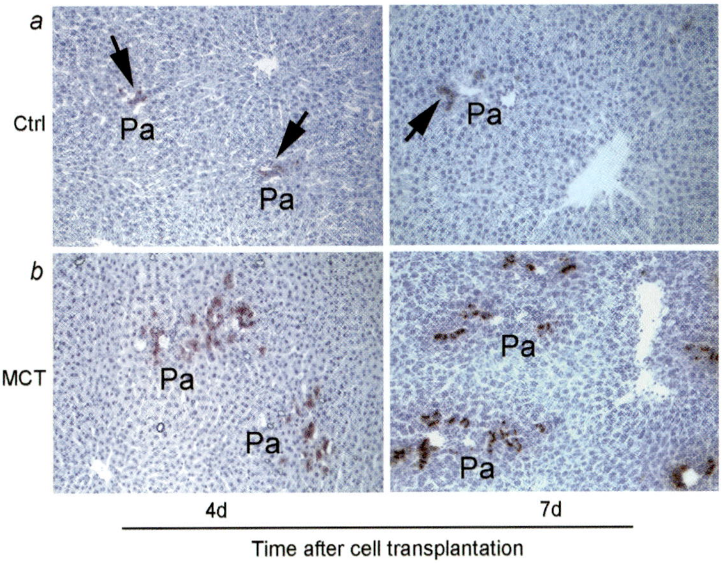

Color Plate 4. Liver preconditioning using monocrotaline (MCT) for improving cell engraftment in DPPIV– rats. Transplanted F344 rat hepatocytes are shown in the recipient liver 4 and 7 days after cell transplantation. Panel **a** shows 1–3 transplanted hepatocytes with histochemically visualized DPPIV activity (red color, arrows) in periportal areas (Pa). By contrast, in MCT-treated rats (**b**) several-fold more transplanted cells are present. Original magnification, ×200; hematoxylin counterstain. Modified from Joseph B, et al. (20). (*see* discussion on p. 111)

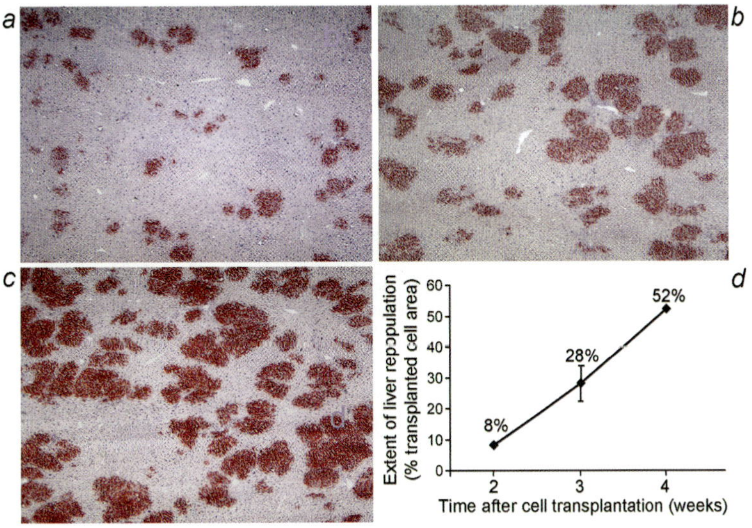

Color Plate 5. Analysis of the kinetics of liver repopulation in DPPIV– rats preconditioned with retrorsine and partial hepatectomy. Foci of transplanted cells with DPPIV activity (red color) are seen 2 (**a**), 3 (**b**), and 4 weeks (**c**) after cell transplantation. Morphometric analysis of liver repopulation in panel **d** indicates linear increase in liver repopulation during this period. Original magnification, (**a**–**c**), ×40; hematoxylin counterstain. Modified from Wu Y-M et al. 18. (*see* discussion on p. 112)

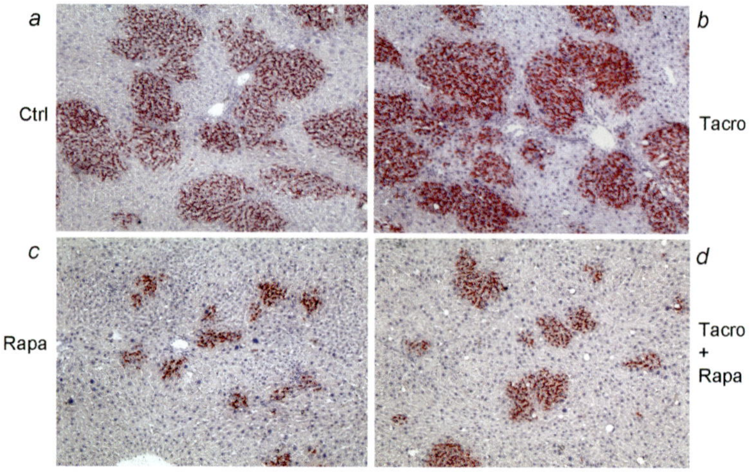

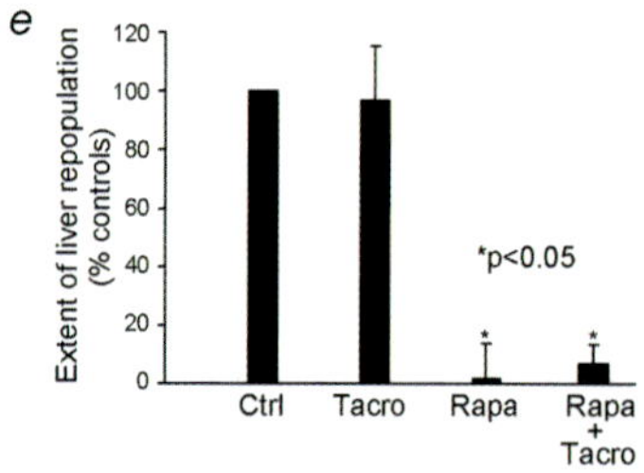

f

Condition (n = 3-5 each)	Size of transplanted cell foci (x10^{-3} mm^{3})	P values vs. controls
Controls	176±12	-
Tacro	169±31	0.99
Rapa	61±8	0.002
Rapa + Tacro	62±7	0.003

Color Plate 6. Effect of immunosuppressive drugs, Rapamycin (Rapa) and Tacrolimus (Tacro), on liver repopulation in DPPIV– rats preconditioned with retrorsine and partial hepatectomy. Animals were treated with drugs subsequent to the completion of cell engraftment. Rapa- but not Tacro-suppressed transplanted cell proliferation as shown by DPPIV histochemistry and morphometric analysis of either the extent of liver repopulation (**e**) or individual transplanted cell foci (**f**). Original magnification (**a–d**), ×100; hematoxylin counterstain. Modified from Wu Y-M et al. (18). (*see* discussion on p. 114)

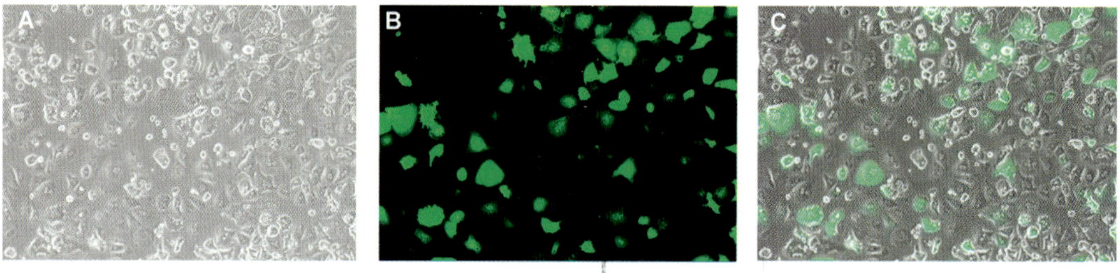

Color Plate 7. Transfection by Amaxa Nucleofection: Expression of GFP in primary mouse hepatocytes (isolated from C57BL/6 mice) nucleofected using an Amaxa mouse hepatocyte Nucleofector kit with a plasmid encoding maxGFP. Twenty-four hours after nucleofection, cells were analyzed by bright field (**A**) and fluorescence microscopy (**B**). The merged image is shown in panel (**C**). (*see* discussion on p. 124)

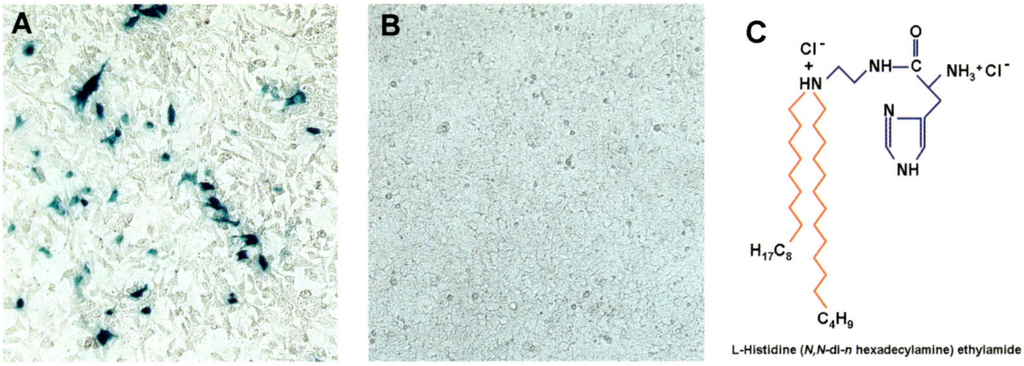

Color Plate 8. Transfection using liposomes containing F protein of the Sendai virus: Expression of LacZ in cells transfected with DNA-loaded F-virosomes as described in the text. After incubation for 24 h, cells were fixed with ethanol, stained for β-galactosidase and photographed. (magnification, ×20, Nikon, Japan). Hepa1 cells (**A**), HEK293 cells (**B**). Note, only asialoglycoprotein-expressed cells are transduced by this method. Structure of histidine lipid used to enhance F-virosome-mediated gene transfer (**C**). (*see* discussion on p. 127)

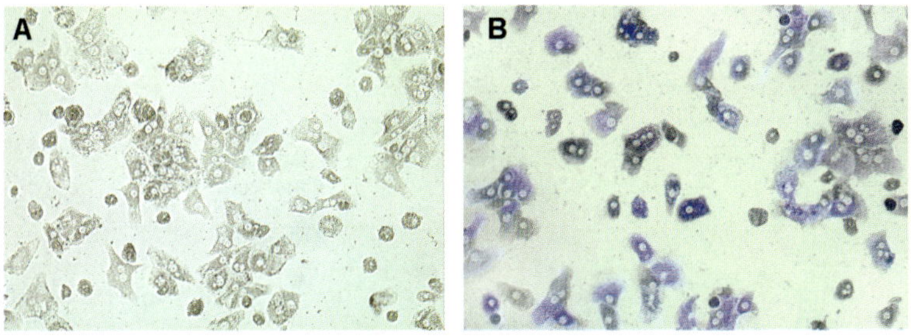

Color Plate 9. Transduction of primary rat hepatocytes using a Lentiviral vector: Isolated Gunn rat hepatocytes were transduced with Lentivirus pAlb-UGT1A1 at an MOI of 10 and immunostained with WP1, monoclonal primary antibody against UGT1A1, followed by anti mouse Alkaline Phosphatase substrate kit III as described in the text and control hepatocytes (**A**) and experimental hepatocytes (**B**) were photographed. (*see* discussion on p. 132)

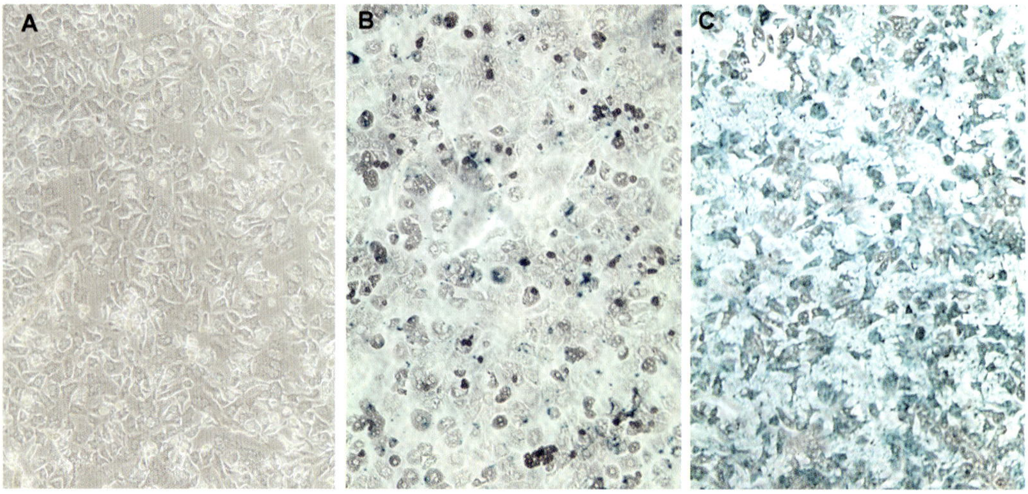

Color Plate 10. Lentiviral vector-mediated transduction of primary mouse hepatocytes, enhanced by Magnetofection®: Isolated mouse primary hepatocytes were transduced with Lentivirus pAlb-LacZ at an MOI of 5 with or without Magnetofection® as described in the text, and were stained 48 h later for bacterial β-galactosidase activity (blue reaction products). (**A**) Untransfected control; (**B**) Lentiviral transduction without Magnetofection®; (**C**) Lentiviral transduction enhanced by Magnetofection®. (*see* discussion on p. 133)

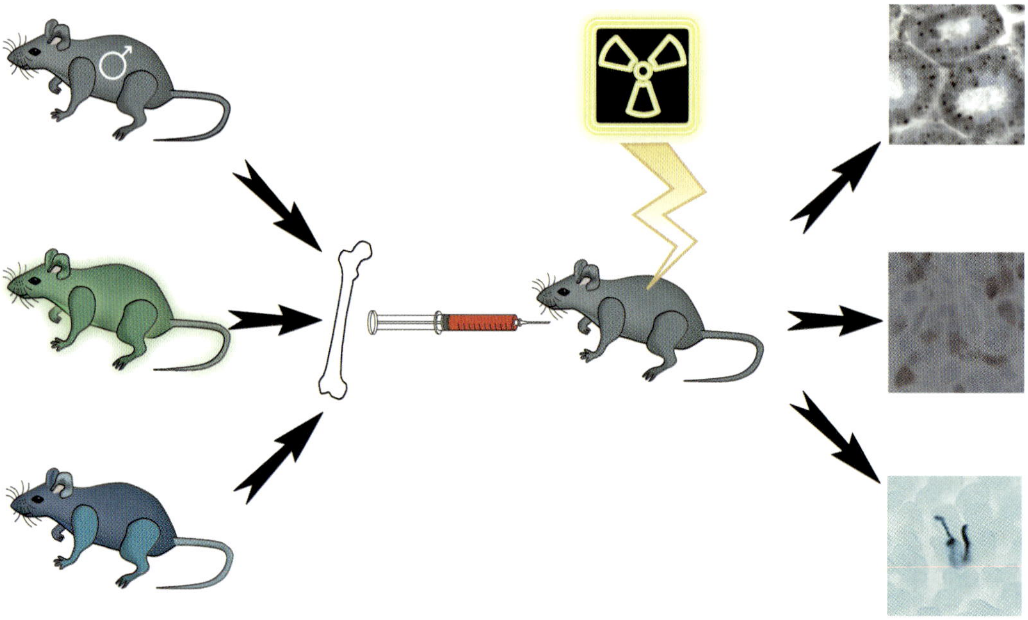

Color Plate 11. Revealing that bone marrow cells (BMCs) have differentiated into non-haematopoietic cells can be achieved by transplanting lethally irradiated animals with new BMCs that can be tracked whatever their subsequent fate. This would include male BMCs to a female recipient, or GFP- or LacZ-positive BMCs to wild-type recipients. The male chromosome can be detected by in situ hybridisation, GFP by immunohistochemistry and β-galactosidase by X-gal histochemistry. (*see* discussion on p. 141)

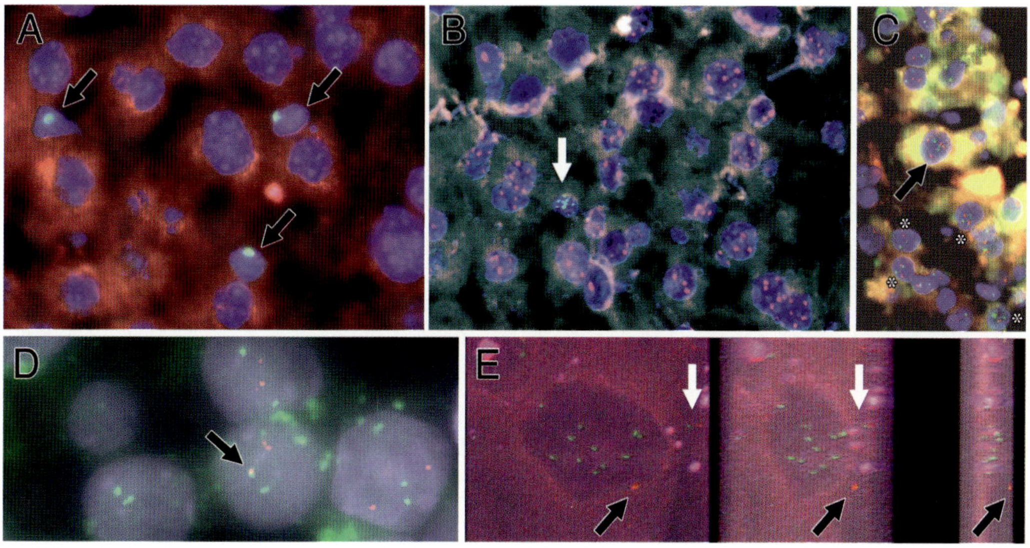

Color Plate 12. Fluorescent and confocal microscopy. (**A**) Male cells (arrows) in male bone marrow-transplanted female mouse liver (green FITC dot). These cells are CK18 immunoreactive (red cytoplasm), suggestive of hepatocyte differentiation. (**B**) Human cell (green FITC, spotty nucleus, arrowed) in mouse liver (pink CY3 spots) after injection of human $CD133^{+}$ cells into a NOD-SCID mouse. (**C**) BCR/ABL probe on human liver in a case of CML showing normal ploidy, with two copies of chromosome 9 (red signals) and two copies of chromosome 22 (green signals) in some cells (asterisks), but multiple copies (polyploidy) in another cell (arrow). (**D**) BCR/ABL fusion signal (green and red overlap producing orange, arrowed) seen in cell tentatively identified as a hepatocyte in a case of CML. There is one native chromosome 9 (red), one native chromosome 22 (green) and one small red signal (*ASS* gene). (**E**) Confocal images demonstrating liver polyploidy in a female mouse transplanted with male bone marrow, with multiple X chromosomes (green signals) showing that a Y chromosome (red signal, black arrow) is outside the nuclear membrane (view **E**), while a smaller nucleus (white arrow) has both X and Y chromosomes contained within it. (*see* discussion on p. 142)

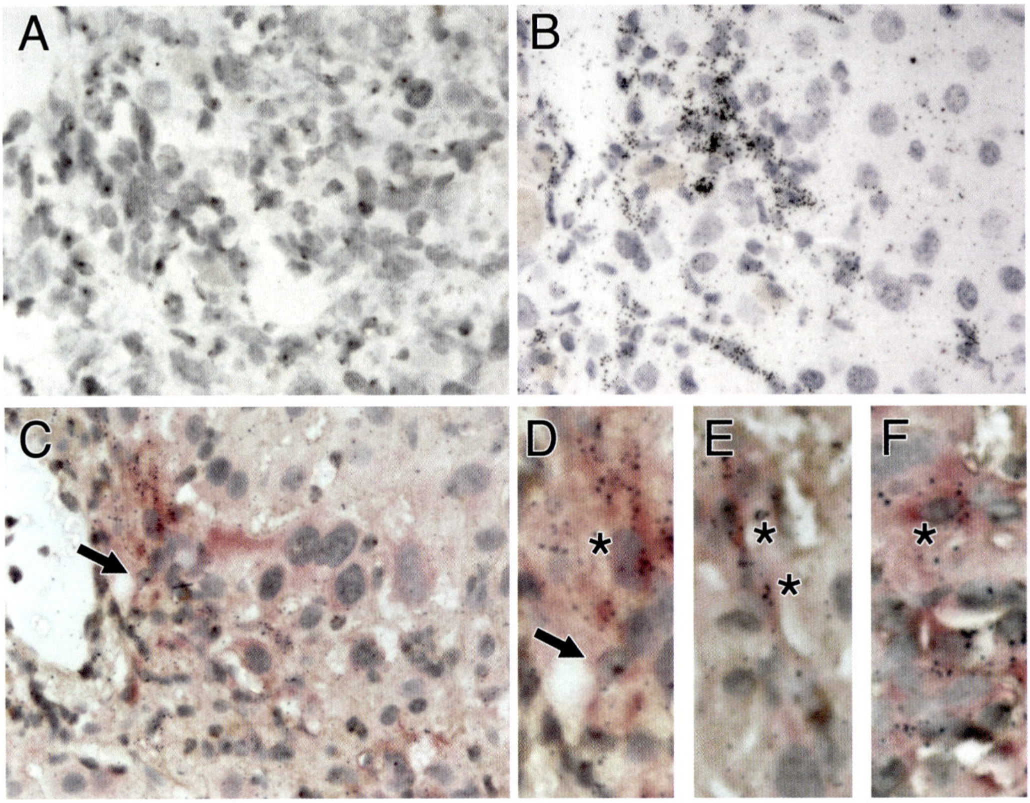

Color Plate 13. Liver fibrosis in a mouse as viewed by bright field microscopy. (**A**) Demonstration of Y chromosome-positive cells (brown nuclear dots) in a female mouse liver after a male bone marrow transplant. (**B**) Demonstration of mRNA for pro(α1)I (black autoradiographic grains) in the same liver using a ^{3}H-labelled antisense riboprobe. (**C**) Demonstration of Y chromosome detection (brown dot, arrow) and IHC for α-SMA expression (red staining) – a marker of myofibroblast differentiation. (**D**) Demonstration of the expression of mRNA for pro(α1)I, the Y chromosome and α-SMA in the same liver. One Y chromosome-positive cell is expressing neither α-SMA nor mRNA for pro(α1)I, but another cell (asterisk) is expressing all three markers. Note the reduced grain density when techniques are combined in comparison to when ISH for the mRNA is performed alone. (**E** and **F**) Examples of ISH for pro(α1)I mRNA expression and immunoreactivity for α-SMA in the same section. (*see* discussion on p. 147)

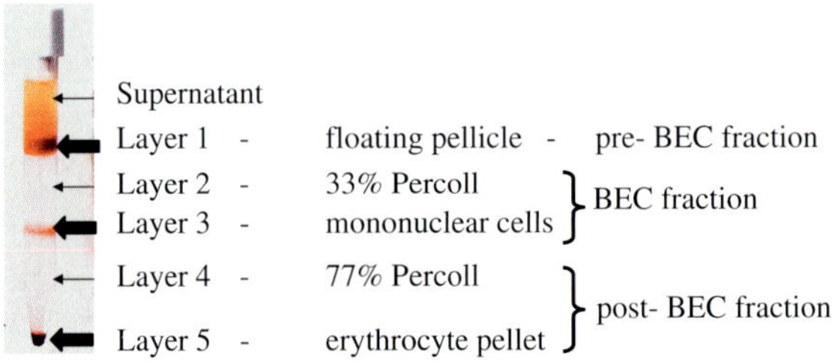

Color Plate 14. The appearance of a Percoll gradient following centrifugation at 800×g for 30 min is shown. Layers 2 and 3 contain biliary epithelial cells (approximately 10%) and are harvested for further purification of immature and mBEC populations by immunomagnetic separation. The supernatant and fractions 1 and 4–6 are discarded. (*see* discussion on p. 197)

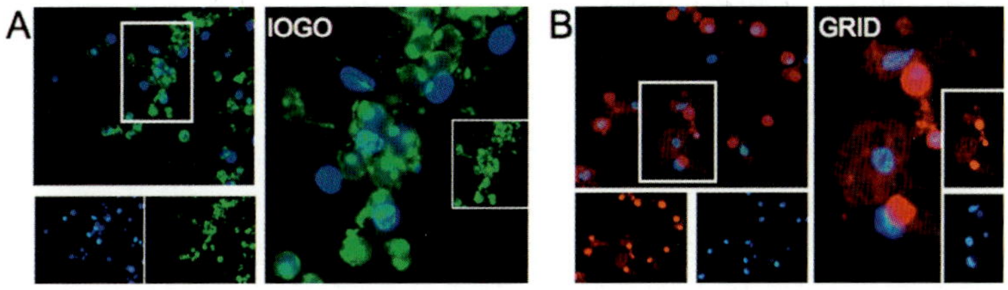

Color Plate 15. Visualisation of the MRI contrast agent. (**A**) Adult human hepatocytes being labelled with the bimodal Iron Oxide Green Oregon (IOGO) contrast agent (in green). Note that some cells (cell nuclei in blue) are not labelled. It is noteworthy that the contrast agent seems strongly associated with the cell nuclei and does not fill the cytoplasm. It is likely that mainly phagocytic Kupffer cells incorporated this agent, whereas unlabelled cells represent a small fraction of undifferentiated hepatocytes. (**B**) In contrast, the Gadolinium Rhodamine Dextran (GRID) bimodal agent (*in red*) clearly labels the cytoplasm of cells that have the appearance of immature hepatocytes and is incorporated into all types of cells. (*see* discussion on p. 212)

activity of the F-protein and lysis of endosomes that releases the DNA into the cytosol. The protocol for generating F-virosome–-DNA complexes, with or without the histidinylated lipid, is given below.

3.4.1. Entrapment of the DNA of Interest into Reconstituted F-Virosomes (22–24)

The protocol described here is from 100 mg of Sendai virus for the preparation of hepatocyte-specific F-virosomes.

1. Spin 100 mg of Sendai virus at 100,000×*g* for 1 h at 4°C.
2. Suspend pellet in 0.02 M Tris-buffered saline (pH 8.3). To this, add 30 mM DTT (Sigma) solution so that the final concentration of DTT is 3 mM.
3. Mix the sample properly and incubate the sample in a 37°C water bath for 4 h with occasional shaking.
4. For dialysis treatment boil the dialysis bag (cutoff 12,000–14,000) in MilliQ water for 5 min.
5. Now thoroughly rinse the bag with cold dialysis buffer (10 mM Tris-buffered saline pH 7.4).
6. Fill the bag with the DTT-treated viral suspension and dialyze the sample against 4 L cold dialysis buffer overnight, giving five changes of 2 h each.
7. After dialysis, spin the sample at 100,000×*g* for 1 h at 4°C.
8. Homogenize the pellet with 10 mM Tris-buffered saline pH 7.4, to this add 10% Triton X-100, double the amount of virus, keeping the final percentage of triton between 2 and 5%.
9. Mix and rotate the sample slowly on a rotator for 1 h at room temperature.
10. Spin the sample at 100,000×*g* for 1 h at 4°C.
11. The histidine-containing lipid was dissolved in solvent (methanol: chloroform, 2:1) and dried in a glass vial under nitrogen to form a thin film (4 mg lipid per 100 mg Sendai virus).
12. To this mixture, add the supernatant from detergent extract containing only the viral F protein and lipids and incubate at 20°C for 30 min with gentle shaking.
13. To this final solution add DNA sample containing 2 mM EDTA and mix.
14. For detergent removal add SM-2 Biobeads (BioRad, eight times the amount of detergent) to the above solution and rotate the sample on a rotator for 2 h at 4°C.
15. Again add the same amount of Biobeads to the above solution and rotate the sample slowly on a rotator for 2 h at room temperature.

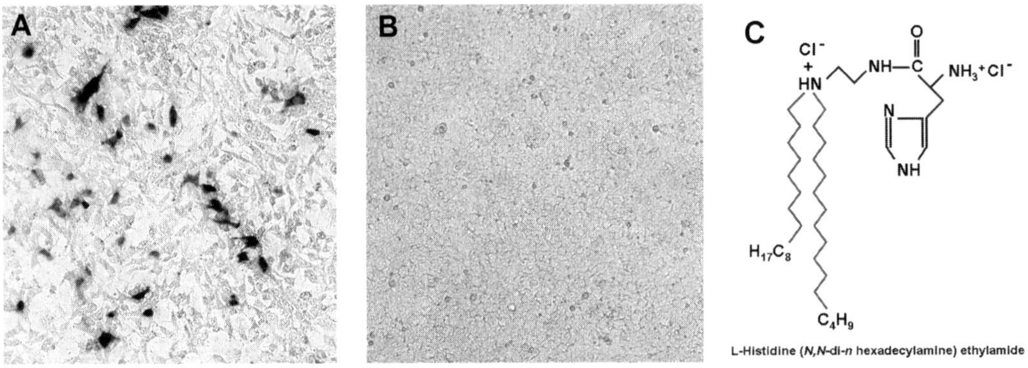

Fig. 11.2. **Transfection using liposomes containing F protein of the Sendai virus:** Expression of LacZ in cells transfected with DNA-loaded F-virosomes as described in the text. After incubation for 24 h, cells were fixed with ethanol, stained for β-galactosidase and photographed. (magnification, ×20, Nikon, Japan). Hepa1 cells (**A**), HEK293 cells (**B**). Note, only asialoglycoprotein-expressed cells are transduced by this method. Structure of histidine lipid used to enhance F-virosome-mediated gene transfer (**C**). (*see* Color Plate 8)

16. Repeat the above step again by adding the same amount of Biobeads at room temperature.
17. Take out the virosome suspension with a 26-gauge needle avoiding Biobeads and spin the suspension at 100,000×*g* for 1 h at 4°C.
18. Wash the pellet in Tris-buffered saline pH 7.4 at 100,000×*g* for 1 h at 4°C.
19. Finally suspend the pellet in 10 mM phosphate buffered saline pH 7.4 and store the virosome samples at 4°C.

3.4.2. Transfection of Primary Hepatocytes with DNA-Loaded F-Virosomes (Fig. 11.2)

1. Primary hepatocytes, isolated by collagenase perfusion, are cultured in DMEM containing 10% FBS, 100 μg/ml streptomycin, 100 U/ml penicillin, 2 mM L-glutamine, 1 mM sodium pyruvate, 25 nmol/L dexamethasone and 5 μg/ml bovine insulin and divided into 100 mm pre-coated plates containing 1.5×10^6 cells at 37°C, 5% CO_2 for 4 h.
2. Primary hepatocytes are rinsed with DPBS and incubated with DNA-loaded modified F-virosomes (2–4 μg per 5×10^5 cells) with serum-free medium for 2 h at 37°C, 5% CO_2.
3. Following the initial incubation, DMEM with 20% FBS is added to the cells, followed by incubation for an additional 24 h.
4. After 24 h of infection, the cells are washed with DPBS. The cells are now ready for analysis of transgene expression or transplantation. For transplantation, the cells are released from the plates by gentle agitation with 0.25% trypsin, harvested by centrifugation at 50×*g* for 5 min at 4°C and resuspended in 0.5 ml PBS .

3.5. Gene Transfer Using Integrating Recombinant Viral Vectors

3.5.1. Ex Vivo Gene Transfer Using Recombinant Oncoretroviruses

Oncoretroviruses, such as Moloney's murine leukemia viruses, have simple genomes that can be readily manipulated for transferring genes into a variety of mammalian cells. Recombinant oncoretroviruses are generated usually in cloned producer cells, whereby the titer and other characteristics of the vector remain similar from batch to batch. It is also possible to generate the vector by transient transfection. However, integration of oncoretroviral vectors requires the host cell to be in the cell cycle, which makes these vectors inefficient for hepatocytes and other quiescent cells. Therefore, increasing numbers of investigators are using lentiviral vectors, which can infect non-dividing cells efficiently (see later in this section). For details of construction of plasmids containing the recombinant cDNA of the retroviral genome and the gene of interest, and for the generation of producer cells and the recombinant virus, see ref. *(25–28)*. The protocol for gene transfer into cultured primary hepatocytes is described in brief below.

1. Hepatocytes are isolated as described above and are cultured on collagen- or gelatin-coated plates for 48–72 h before infection. The culture medium can be DMEM containing 10% FBS, 100 μg/ml streptomycin, 100 U/ml penicillin, 2 mM L-glutamine, 1 mM sodium pyruvate, 25 nmol/l dexamethasone and 5 μg/ml bovine insulin and divided into 100 mm pre-coated plates containing 1.5×10^6 cells for 4 h. To stimulate mitosis of the cultured hepatocytes, some investigators have used hormonally defined media, containing HGF and EGF *(29, 30)*.
2. Hepatocytes are rinsed in DPBS and incubated with recombinant retrovirus at MOI=10 for 2–6 h at 37°C. The medium contains 8 μg/ml polybrene.
3. After incubation, the medium is replaced with complete DMEM and the cells are cultured overnight.
4. The cells are then washed with DPBS and detached from the plate using 0.25% trypsin for 5 min. The cells are collected by centrifugation at $50 \times g$ for 5 min at 4°C and the cell pellet is resuspended in 500 μl PBS for transplantation.

3.5.2. Ex Vivo Gene Transfer Using Recombinant Lentiviruses

As mentioned above, recombinant lentiviruses are particularly attractive for ex vivo gene transfer into hepatocytes because these vectors can infect non-dividing cells efficiently. Furthermore, infection with these vectors occurs rapidly enough for gene transfer into hepatocytes in suspension. This is particular useful for ex vivo gene therapy, because there can be significant loss of cells during detachment of the cells, once they are attached to culture plates. Various generations of lentiviral vector production systems are available. In our laboratory, we have found the 4-plasmid system developed by Naldini and associates to be quite efficient *(31)*. As the vectors are pseudotyped with the vesicular stomatitis G (VSV-G) envelope, they infect a wide variety of cell types. These self-inactivating vectors express the gene of interest from internal promoters. To obtain transcriptional specificity for hepatocytes, a hepatocyte-specific promoter may be employed (e.g. the albumin promoter-enhancer). For ubiquitous transgene expression, other promoters (e.g. the phosphoglycerate kinase promoter) can be used. The example given below is based on the use of the albumin promoter-enhancer. The system is based on the transfection of four plasmids: (i) The transduction plasmid, containing the internal promoter and the coding region of the gene of interest followed by a woodchuck post-transcriptional regulatory element (WPRE) (e.g. pAlb-UGT1A1); (ii) pMD2-VSV-G, the plasmid expressing VAS-G envelope protein; (iii) the core-packaging plasmid, pMDLg/pRRE; and (iv) pRSV-REV, a plasmid that express the REV protein from a Rous sarcoma virus (RSV) promoter.

3.5.2.1. Generation and Concentration of Lentivirus pAlb-UGT1A1

1. The day before transfection, 293T cells were plated at 1.8×10^7 cells per 150 mm dish. Each culture dish contains 22 ml DMEM (Invitrogen) containing 10% FBS, 100 μg/ml streptomycin and 100 U/ml penicillin. Usually, 14 plates are seeded and incubated at 37°C in air containing 5% CO_2.
2. Two hours before transfection, the medium was changed to IMDM (Invitrogen) containing 10% FBS, 100 μg/ml streptomycin and 100 U/ml penicillin.
3. For each 150 mm dish, the following plasmid DNAs are mixed: pMD2-VSV-G 6 μg, pMDLg/pRRE 10 μg, pRSV-REV 5 μg and 32 μg of pAlb-UGT1A1 in a final volume of 900 μl of 0.1× TE/ddH_2O. To this mixture is added 100 μl of 2.5 M$CaCl_2$. The DNA mixture is kept at room temperature for 5 min.
4. After this, add 1 ml of 2× BBS (150 mM NaCl; 50 mM Bes; and 1.5 mM Na_2HPO_4, pH 6.95). The pH of 2× BBS is checked carefully (*see* **Note 7**); the solution is sterilized by filtration (0.22 μm) and stored in 25 ml aliquots at –20°C. Before use, the solution is mixed in a Vortex and incubated for 30 min at room temperature.
5. To each cell culture dish is added dropwise 2 ml of the plasmid DNA solution, while swirling gently to distribute the solution evenly. The cells are incubated overnight at 37°C, in a 5% CO_2 atmosphere.
6. Fourteen to 16 h after transfection, the medium is replaced by fresh medium.

7. Medium is collected after 24 and 48 h, and filtered through a 0.22-μm filter (Millipore, USA).
8. To sediment the recombinant Lentiviral vector, the filtered medium is centrifuged at 19,500 r.p.m. for 2 h at 22°C. The supernatant is discarded carefully and the pellet is resuspended in 500 μl of phosphate buffered saline containing 10 mM phosphate and 150 mM sodium chloride (1× PBS), pH 7.4 and stored at 4°C .
9. If further concentration of the vector is required, the concentrated virus harvested from step 8 is diluted 1:1 with 1× PBS and centrifuged at 19,500 r.p.m. for 2 h at 22°C. The final pellet is resuspended in 500 μl of the above buffer, aliquoted and stored at –80°C till further use (*see* **Note 8**).
10. Determination of viral titer: In cases where the vector expresses a marker gene or a gene product that is easily visualized by cytochemical or immunocytochemical staining, the titer is determined from the maximum dilution resulting in positively staining cells after 72 h. For vectors that do not express a gene product that can be readily visualized, viral titer can be determined by quantitative RT-PCR for WPRE. Examples of titer determination methods in three different scenarios are given below.
    a) For transfer vector expressing a marker gene, e.g. lentivirus pAlb-LacZ, 72 h after infection with the various dilutions of the vector, enzyme-cytochemical staining was performed to determine β-galactosidase expression. Briefly, the transduced cells were fixed for 5 min with 100% ethanol, washed with PBS twice, incubated with the complete X-gal staining solution (1 mg/ml X-Gal, 35 mM potassium ferricyanide, 35 mM potassium ferrocyanide, 2 mM $MgCl_2$) at 37°C for 1 h to overnight. The viral titer was calculated as follows:

       Transduction Units (TU)/ml=[Numbers of transduced cells (10^5)]×[% of positive stained cells]×Dilution factor/100
    b) For pAlb-UGT1A1, which expresses the human UGT1A1, viral gene transfer titer was determined by immunocytochemical probing with an antibody that is specific for the transgene product (e.g. WP1, a monoclonal antibody against the human UGT1A family of proteins) *(32)*. In this case, mouse hepatoma cells (Hepa-I) were used for tittering. The transduced Hepa-I cells were washed twice with 10 mM PBS and fixed for 45 min with acetone/ methanol (1:4) at room temperature. The fixed cells were washed with 70% ethanol in PBS and three times with Phosphate buffer containing 0.5% Tween-20 and 1% BSA (buffer A). The cells were incubated with WP1 for 45 min at room temperature and then washed five times with

buffer A. The goat anti-mouse IgG conjugated with AKP, used as secondary antibody, was detected by Alkaline Phosphatase substrate kit III (Vector laboratories, USA). The viral titer was calculated as follows:

Transduction Units (TU)/ml = [Numbers of transduced cells (10^5)] X [% of positive stained cells] X Dilution factor /100

c) For transfer vectors that do not express easily visualized gene products, the copy number of WPRE, which is part of 3′ UTR of many lentiviral transgene mRNA transcripts, was used to determine the virus titer *(33)*, using qRT-PCR. The total RNA was isolated from the transduced Hepa-I cells, using RNeasy columns as per the manufacturer's guidelines (Qiagen, Germany). The 1 μg purified RNA was reverse transcribed to generate cDNA as usual. The sense primer (1277F), 5′-CCGTTGTCAGG-CAACGTG-3′, antisense primer (1361R), 5′-AGCTGACAGGTGGTGGCAAT3′, probe (1314P) and 5′-FAM-TGCTGACGCAACCCCCACTGGT-TAMRA-3′ were used to detect the WPRE sequence. Expression levels of WPRE were determined by qRT-PCR, and β-actin mRNA was used as normalization. The primers and probe of β-actin were as follows *(34)*: forward primer, 5′-TCACCCACACTGTGCCCATCTACGA-3′ reverse primer: 5′-GGATGCCACAGGATTCCATACCCA-3′; probe 5′-FAM-TATGCTCTCCCTCACGCCATCCTGCGT-TAMRA-3′. The transfer vector was used to generate the standard curve, viral titer was determined as follows: Transduction Units (TU)/ml=[numbers of WPRE molecules in transduced cells (10^5)]×Dilution factor/100.

3.5.2.2. Transduction of Primary Hepatocytes Subsequent to Transplantation (Fig. 11.3)

1. Isolate rat or mouse primary hepatocytes by liver perfusion using collagenase as above. Hepatocytes from other species, such as rabbit or human, can also be processed by the following method.
2. Cells are washed twice by sedimenting at 50×*g* for 5 min and resuspending (10^7cells/ml) in DMEM containing 10% FBS, 100 μg/ml streptomycin, 100 U/ml penicillin, 2 mM L-glutamine, 25 nmol/L dexamethasone and 5 μg/ml bovine insulin. Cell viability is determined by trypan blue exclusion.
3. Hepatocytes are incubated for 4 h at 4°C with the recombinant vector (e.g. pAlb-LacZ at MOI=10) in the presence of 8 μg/ml polybrene. *(35, 36)* Some investigators prefer centrifuging the mixture at room temperature for 4 h at 50×*g* to enhance gene transfer *(37)*. In our hands, this procedure did not increase gene transfer over simply incubating the cells with the vector.

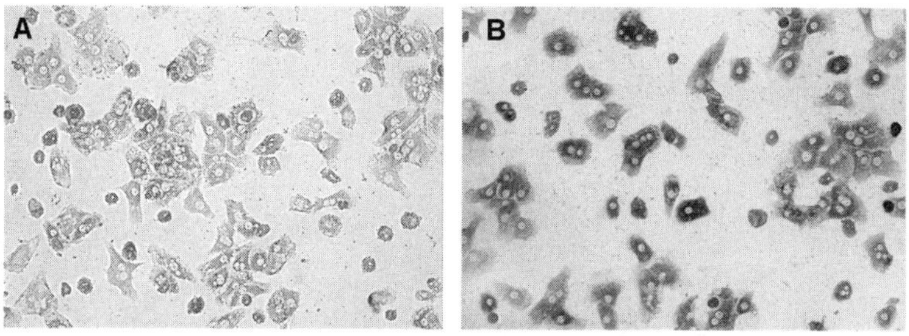

Fig. 11.3. **Transduction of primary rat hepatocytes using a Lentiviral vector:** Isolated Gunn rat hepatocytes were transduced with Lentivirus pAlb-UGT1A1 at an MOI of 10 and immunostained with WP1, monoclonal primary antibody against UGT1A1, followed by anti mouse Alkaline Phosphatase substrate kit III as described in the text and control hepatocytes (**A**) and experimental hepatocytes (**B**) were photographed. (*see* Color Plate 9)

4. After 4 h of incubation at 4°C, the cells are incubated at 37°C for 15 min, washed twice, resuspended in PBS and then used for transplantation.
5. For retrospective testing of the transduction efficiency of lentiviral vector in vitro, transduced cells are plated at 1×10^5/well of 24-well plates pre-coated with bovine dermal collagen (Vitrogen, Cohesion Technologies). After 48 h, the cells are washed twice with PBS, fixed with 100% ethanol for 5 min and then examined for expression of the marker gene.

3.5.2.3. Enhancement of Gene Transfer by Application of a Magnetic Field (Magnetofection®)

This method is based on magnetic nanoparticles, coated with cationic molecules that permit the particles to be associated with recombinant viral vectors or naked plasmid DNA. Exertion of magnetic force on the particles results in concentration of the nanoparticle–vector complex on cell surfaces, resulting in increased transduction efficiency (**Fig. 11.4**). The variants of these magnetic particles that are designed to work with adenoviral and retroviral (including Lentiviral) vectors are named ViroMag® and ViroMag R/L® by their manufacturers (Oz Biosciences). The protocol for use with one lentiviral vector is given below as an example.

1. Plate the primary hepatocytes 14–16 h before infection in 100 mm tissue culture dishes.
2. Add 150 µl of **ViroMag R/L** in a tube large enough to contain the volume of virus preparation (~0.8 ml). If required, **ViroMag R/L** can be diluted with deionized water (not medium). Medium containing retroviral or lentiviral vectors can be used directly, or diluted as needed with the hepatocyte culture medium.
3. Add the Lentiviral preparation (e.g. pAlb-LacZ) to the tube(s) containing **ViroMag R/L** and mix immediately by pipetting up and down. Incubate for 15 min at room temperature.

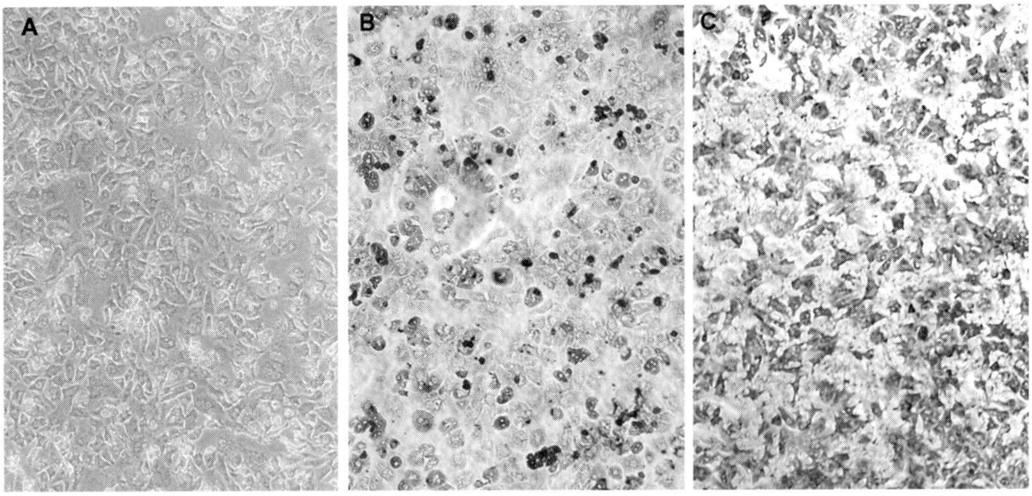

Fig. 11.4. **Lentiviral vector-mediated transduction of primary mouse hepatocytes, enhanced by Magnetofection®:** Isolated mouse primary hepatocytes were transduced with Lentivirus pAlb-LacZ at an MOI of 5 with or without Magnetofection® as described in the text, and were stained 48 h later for bacterial β-galactosidase activity (blue reaction products). (**A**) Untransfected control; (**B**) Lentiviral transduction without Magnetofection®; (**C**) Lentiviral transduction enhanced by Magnetofection®. (*see* Color Plate 10)

4. Add the **ViroMag R/L**-virus mixture to the medium (final volume 8 ml/plate) and add to the plate containing the attached hepatocytes.
5. Place the cell culture plate on the magnetic plate for 15–60 min. Change the medium and culture the cells under standard conditions.
6. The cells can be evaluated for transgene expression after a desired length of time, or released from the plate using trypsin/EDTA immediately after infection for transplantation.

3.6. Gene Transfer Using Non-integrating Recombinant Viral Vectors

Viral vectors that do not integrate into host genomes are lost upon cell division. Therefore, they are not generally employed for ex vivo gene therapy applications involving repopulation of the liver with transplanted cells or for long-tern gene therapy. Recombinant adenoviral vectors and adenoassociated viral vectors fall in this category. For some specific applications, however, adenovectors could be useful. For example, immediate evaluation of the proportion of transplanted cells that engraft and for determining the distribution of the transplanted cells, short-term expression of a marker gene should be sufficient. Adenoviral vectors are particularly advantageous for this type of application, because of their high transduction efficiency toward hepatocytes and the ease of production of the vector at high titers. Reagent kits for several different methods for generating adenoviral vectors are available commercially. One protocol for generation of a replication-incompetent adenoviral vector and ex vivo gene transfer into hepatocytes is described below. In this example, the gene of interest, *LacZ* (expressing *E. coli* β-galactosidase), is inserted into the adenoviral plasmid, using the Adeno-X Expression system (BD Biosciences Clontech).

3.6.1. Generation, Amplification and Purification of Recombinant Adenoviral Vectors

1. HEK293 cells(Note) are plated at 1×10^6 cells in six-well plates containing 20 ml DMEM, supplemented with 10% FBS (Sigma), 100 μg/ml streptomycin and 100 U/ml penicillin, and incubated at 37°C under 5% CO_2 for 24 h before transfection.
2. The adenoviral plasmid vector is digested with *PacI* and transfected into HEK293 cells at 80% confluence, using Lipofectamine.
3. After 24 h, and periodically thereafter, the cells are examined for cytopathic effect (CPE).
4. Once CPE appears, the cells are harvested and ruptured by three freeze–thaw cycles to release the recombinant adenovirus.
5. To amplify the recombinant adenovirus, fresh HEK293 in six-well plates are infected by adding cell lysate from step 4.
6. Repeat step 4. At this point, the cell lysate should be examined for transgene expression by Western blot (or any other method that is available for the transgene of interest).
7. For further amplification, HEK293 cells are plated in three-tier flasks (total surface area 500 cm^2), and grown to 50–80% confluency.
8. The medium is replaced with 90 ml of medium that contains 2% FBS and cell lysate from step 6. The cells are incubated for 24–72 h at 37°C under 5% CO_2 until CPE appears.
9. The infected HEK293 cells are sedimented by centrifugation for 10 min at $650 \times g$ (GSA, 2000 r.p.m.). The cells are resuspended in 10 mM Tris at pH 8.0.
10. The viral particles are purified by $CsCl_2$ gradient centrifugation as follows. One volume of Freon (Fisher) is added to the resuspended cells, shaken for 5 min, centrifuged for 5 min at 2000 r.p.m., 4°C. The upper layer is saved and the lower layer is re-extracted with the addition of 7 ml 10 mM Tris, The first and second upper layers are combined and re-extracted one more time with Freon. The gradient maker is loaded with 10 ml 1.2p CsCl in the distal well and 10 ml 1.4p CsCl in the proximal well. Add the infected cell extract gently against the side of the tube over the gradient and ultracentrifuge at 22,000 r.p.m., 4°C, using SW28 rotor with buckets overnight. The band containing the virus is collected after ultracentrifugation, diluted with one volume of 10 mM Tris,

loaded to the same gradient as above and centrifuged for 4 h at 22,000 r.p.m., 4°C, using SW28 rotor with buckets. The band containing the virus is collected from the tube. The virus particle can be measured by absorbance at 260 nm. Ten microliters of the virus is diluted with 1 ml 1% SDS and OD260 nm is measured 15 min after adding virus to the SDS, using 1% SDS as a blank. The following formula is used to calculate the viral particle: Total viral particles = OD260 × (virus dilution factor) × (0.28 mg/OD) × (3.1×10^{12} particles/mg).

11. The virus preparation is mixed with an equal volume of glycerol and stored at –20°C.
12. The titer of recombinant adenovirus is determined by plaque assay as follows:
    5×10^5 HEK 293 cells are seeded in 12-well plates with growth medium, 100 μl of 10-fold serial-diluted virus is made from 10^{-2} to 10^{-6} ml, is added to each well and incubated at 37°C in 5% CO_2 for 48 h. The cells are fixed with ice-cold 100% methanol at –20°C for 10 min, washed with 1% BSA in PBS three times, incubated with 1:1000 mouse anti-hexon antibody (BD) for 1 h at 37°C, washed with 1% BSA in PBS three times, incubated with 1:500 rat anti-mouse antibody (HRP conjugate, BD) for 1 h at 37°C, washed with 1% BSA in PBS three times and incubated with DAB substrate (BD) solution at room temperature for 10 min. The DAB is aspirated from plate and 1 x PBS is added to the cells. The minimum of three fields of positive cells are counted using a microscope with a 20 X objective, the infectious units for each well is calculated as follows:
    PFU = (infected cells/field) × (fields/well)/volume virus (ml) × (dilution factor).

3.6.2. Transduction of Hepatocytes or Hepatocyte Progenitor Cells

Primary hepatocytes are prepared by collagenase perfusion as described above. Adenoviral vectors can also be used for genetic marking of hepatic progenitor/stem cells. The procedure for transducing primary hepatocytes is described below.

1. Primary hepatocytes are cultured in complete DMEM (DMEM containing 10% FBS, 100 μg/ml streptomycin, 100 U/ml penicillin, 2 mM L-glutamine, 1 mM sodium pyruvate, 25 nM dexamethasone, and 5 μg/ml bovine insulin) and plated on 100 mm Primaria™ plates, coated with collagen (Vitrogen) (1.5×10^6 cells per plate). The cells were allowed to attach to the plate for 4 h.
2. Primary hepatocytes are rinsed in PBS and incubated in DMEM with the recombinant adenovirus (e.g. Ad-LacZ) at a multiplicity of infection (MOI) of 10 in serum-free DMEM for 3 h at 37°C.

3. After the initial incubation, the complete DMEM (as in step 1) is added and the cells are incubated for an additional 24 h.
4. The cells can be detached from the plates by 5-min incubation with 0.05% trypsin in 0.53 mM EDTA for 5 min. The cells are collected by centrifugation at $50\times g$ for 5 min and resuspended in 500 μl PBS. Cell viability is determined by trypan blue exclusion before transplantation.
5. To determine the transduction efficiency, the cells are cultured for an additional 24 h before staining for β-galactosidase expression as described above.

4. Notes

1. Collagen is prepared at 1:10 in 0.012 HCl, coated cell culture plates at room temperature for overnight. The coated plates are transferred to 4°C and washed with 1× PBS twice before using.
2. Nucleofector Solution are stable for 3 months at 4°C after the supplement is added to Nucleofector and pre-warmed to room temperature before using.
3. DTT is an unusually strong reducing agent and liable to air oxidation, so always keep DTT at 4°C and make fresh DTT solution.
4. Viromag R/L could not be frozen, always kept at 4°C and only diluted with deionized water if needed.
5. Freon should be ice-cold before using because cold Freon gas can delipidate the infected cells and extract the virus from cells.
6. Transgene expression can be detected within 6 h after Nucleofection. The rapid expression of the transgene is presumably due to the more efficient transfer of the DNA to the nucleus than with other methods of gene transfer.
7. pH of 2× BBS is very important for transfection efficiency, so should carefully adjust pH of 2× BBS to 6.95.
8. The reason for making multiple aliquots of the virus stocks is to prevent virus titer to decrease by the freeze–thaw cycle.

Acknowledgments

This work was supported by the NIH grants: RO1 DK46057 (to JRC), RO1 DK067440 (to JRC); RO1 DK-068216-02 (to JRC) and RO1 DK039137 (to NRC); by the Gene Therapy Core of the

Human Genetics Program of the Albert Einstein College of Medicine, and a research grant provided by the National Research Development Corporation of India.

References

1. Roy-Chowdhury, J., Grossman, M., Gupta, S., et al. (1991) Long-term improvement of hypercholesterolemia after ex vivo gene therapy in LDLR-deficient rabbits. *Science* 254, 1802–1805.
2. Grossman, M., Raper, S. E., Kozarsky, K., et al. (1994) Successful ex vivo gene therapy directed to liver in a patient with familial hypercholesterolaemia. *Nat Genet* 6, 335–341.
3. Shah, K., Jacobs, A., Breakefield, X. O., et al. (2004) Molecular imaging of gene therapy for cancer. *Gene Ther* 11, 1175–1187.
4. Weissleder, R., Ntziachristos, V. (2003) Shedding light onto live molecular targets. *Nat Med* 9, 123–128.
5. Blasberg, R. (2002) PET imaging of gene expression. *Eur J Cancer* 38, 2137–2146.
6. Landis, C. S., Yamanouchi, K., Zhou, H., et al. (2006) Noninvasive evaluation of liver repopulation by transplanted hepatocytes using 31P MRS imaging in mice. *Hepatology* 44, 1250–1258.
7. Nakamura, J., Okamoto, T., Schumacher, I. K., et al. (1997) Treatment of surgically induced acute liver failure by transplantation of conditionally immortalized hepatocytes. *Transplantation* 63, 1541–1547.
8. Schumacher, I. K., Okamoto, T., Kim, B. H., et al. (1962) Transplantation of conditionally immortalized hepatocytes to treat hepatic encephalopathy. *Hepatology* 24, 337–343.
9. Mashalova, E. V., Guha, C., Roy-Chowdhury, N., et al. (2007) Prevention of hepatocyte allograft rejection in rats by transferring adenoviral early region 3 genes into donor cells. *Hepatology* 45, 755–766
10. Kren, B. T., Ghoss, S. S., Linchen, C. L., et al. (2003) Hepatocyte-targeted delivery of Sleeping Beauty mediates efficient gene transfer in vivo. *Gene Ther Mol Biol* 7, 229–238
11. Ivics, Z., Hackett, P. B., Plasterk, R. H., et al. (1997) Molecular reconstruction of Sleeping Beauty, a Tc1-like transposon from fish, and its transposition in human cells. *Cell* 91, 501–510.
12. Mikkelsen, J. G., Yant, S. R., Meuse, L., et al. (2003) Helper-Independent Sleeping Beauty transposon–transposase vectors for efficient nonviral gene delivery and persistent gene expression in vivo. *Mol Ther* 8, 654–665.
13. Markwell, M. A., Portner, A., Schwartz, A. L. (1995) An alternative route of infection for viruses: entry by means of the asialoglycoprotein receptor of a Sendai virus mutant lacking its attachment protein. *Proc Natl Acad Sci USA* 82, 978–982.
14. Neufeld, D. S. (1997) Isolation of rat liver hepatocytes. *Methods Mol Biol* 75 145–151.
15. Berry, M. N., Friend, D. S. (1969) High-yield preparation of isolated rat liver parenchymal cells: a biochemical and fine structural study. *J Cell Biol* 43, 506–520.
16. Wilson, J. M., Roy-Chowdhury, N., Grossman, M., et al. (1991) Transplantation of allogeneic hepatocytes into LDL-receptor deficient rabbits leads to transient improvement in hypercholesterolemia. *Clin Biotechnol* 31, 21–26
17. Sen, L., Gambhir, S. S., Furukawa, H., et al. (2005) Noninvasive imaging of ex vivo intracoronarily delivered nonviral therapeutic transgene expression in heart. *Mol Ther* 12, 49–57.
18. Sen, L., Hong, Y. S., Luo, H., et al. (2001) Efficiency, efficacy and adverse effects of adenovirus versus liposome-mediated gene therapy in cardiac allografts. *Am J Physiol* 281, H1433–H1441.
19. Oshima, K., et al. (2002) Localized interleukin-10 gene transfer induces apoptosis of alloreactive T cells via Fas/FasL pathway, improves function and prolongs survival of cardiac allografts. *Transplantation* 74, 1019–1026.
20. Daftary, G. S., Taylor, H. S. (2001) Efficient liposome-mediated gene transfection and expression in the intact human uterus. *Human Gene Therapy* 12, 2121–2127.
21. Aslan, H., Zilberman, Y., Arbeli, V., et al. (2006) Nucleofection-based ex vivo nonviral gene delivery to human stem cells as a platform for tissue regeneration. *Tissue Eng* 12, 877–889.

22. Bagai, S., Puri, A., Blumenthal, R., et al. (1993) Hemagglutinin-neuraminidase enhances F protein-mediated membrane fusion of reconstituted Sendai virus envelopes with cells. *J Virol* 7, 3312–3318.
23. Bagai, S., Sarkar, D. P. (1993) Reconstituted Sendai virus envelopes as biological carriers: dual role of F protein in binding and fusion with liver cells. *Biochim Biophys Acta* 1152, 15–25.
24. Santosh, K. V., Mani, P., Sharma, N. R., et al. (2005) Histidylated lipid-modified Sendai viral envelopes mediate enhanced membrane fusion and potentiate targeted gene delivery. *J Biol Chem* 280, 35399–35409.
25. Blesch, A. (2004) Lentiviral and MLV based retroviral vectors for ex vivo and in vivo gene transfer. *Methods* 33, 164–172
26. Danos, O., Mulligan, R. C. (1988) Safe and efficient generation of recombinant retroviruses with amphotropic and ecotropic host ranges. *Proc Natl Acad Sci USA* 85, 6460–6464.
27. Pear, W. S., Nolan, G. P., Scott, M. L., et al. (1993) Production of high-titer helper-free retroviruses by transient transfection. *Proc Natl Acad Sci USA* 90, 8392–8396.
28. Roy-Chowdhury, J., Grossman, M., Gupta, S., et al. (1991) Long term improvement of hypercholesterolemia after ex vivo gene therapy in LDL-receptor deficient rabbits. *Science* 254, 1802–1805.
29. Minami, H., Tada, K., Roy Chowdhury, N., et al. (2000) Enhancement of retrovirus-mediated gene transfer to rat liver in vivo by infusion of hepatocyte growth factor and triiodothyronine. *J Hepatol* 33, 183–188.
30. Bosch, A., McCray, P. B. Jr., Walters, K. S., et al. (1998) Effects of keratinocyte and hepatocyte growth factor in vivo: implications for retrovirus-mediated gene transfer to liver. *Hum Gene Ther* 9, 1747–1754.
31. Follenzi, A., Naldini, L. (2002) Generation of HIV-1 derived lentiviral vectors. *Methods Enzymol* 346, 454–465.
32. Seppen, J., Tada, K., Hellwig, S., et al. (1996) Bilirubin glucuronidation by intact Gunn rat fibroblasts expressing bilirubin UDP-glucuronosyltransferase *Biochem J* 314, 477–483.
33. Lizee G, Aerts JL, Gonzales MI, Chinnasamy N, Morgan RA, Topalian SL. (2003) Real-time quantitative reverse transcriptase-polymerase chain reaction as a method for determining lentiviral vector titers and measuring transgene expression. *Hum Gene Ther* 14, 497–507
34. Pahl A, Kühlbrandt U, Brune K, Röllinghoff M, Gessner, A. (1999) Quantitative detection of *Borrelia burgdorferi* by Real-Time PCR. *J Clin Microbiol* 37, 1958–1963.
35. Giannini C, Morosan S, Tralhao JG, Guidotti JE, Battaglia S, Mollier K, Hannoun L, Kremsdorf D, Gilgenkrantz H, Charneau P.A (2003) Highly efficient, stable, and rapid approach for ex vivo human liver gene therapy via a FLAP lentiviral vector. *Hepatology* 38, 114–122.
36. Nguyen TH, Oberholzer J, Birraux J, Majno P. Morel P, Trono D. (2002) Highly efficient lentiviral vector-mediated transduction of nondividing, fully reimplantable primary hepatocytes. *Mol Ther* 6, 199–209.
37. Cui Y, Chang LJ. (2003) Detection and selection of Lentiviral vector-transduced cells. *Methods in Molecular Biology* 229, 69 85.

Chapter 12

Sources of Adult Hepatic Stem Cells: Haematopoietic

Rosemary Jeffery, Richard Poulsom, and Malcolm R. Alison

Abstract

Bone marrow cells can engraft in the liver and differentiate into a variety of cell types including hepatocytes and myofibroblasts. This chapter describes how, after transplantation of male bone marrow into female recipients, cells of bone marrow origin (male) can be identified in the female liver by virtue of detection of the Y chromosome by the technique of in situ hybridisation (ISH). Furthermore, ISH for Y chromosome detection can be combined both with immunohistochemistry (IHC) to identify phenotype and with ISH for mRNA to demonstrate function. Additionally, we show that bone marrow-derived cells can be identified in the liver without prior sex-mismatch bone marrow transplantation, identifying instead the *BCR:ABL* fusion gene that is present in all such cells in almost all patients suffering from chronic myelogenous leukaemia (CML).

Keywords: Bone marrow, X and Y chromosomes, in situ hybridization, immunohistochemistry, myofibroblast, chronic myelogenous leukaemia, *BCR:ABL* gene.

1. Introduction

Our ability to track and identify cells from outside the liver that are able to act as progenitors for liver cells, in particular for hepatocytes, relies on identifying the origin of these so-called ‘plastic cells’ and characterising their phenotype. This may be achieved in a variety of ways. In experimental models, one of the simplest approaches is to lethally irradiate female mice, rescue them with a male bone marrow transplant and look for Y chromosome-expressing donor cells within the liver or other organs of interest. Other approaches rely on a similar replacement of recipient bone marrow with cells that carry other markers such as green fluorescent protein (GFP), Luciferase or the *Escherichia coli* β-galactosidase gene (**Fig. 12.1**)

Anil Dhawan, Robin D.Hughes (eds.), *Hepatocyte Transplantation, vol. 481*

DOI 10.1007/978-1-59745-201-4_12 Springerprotocols.com

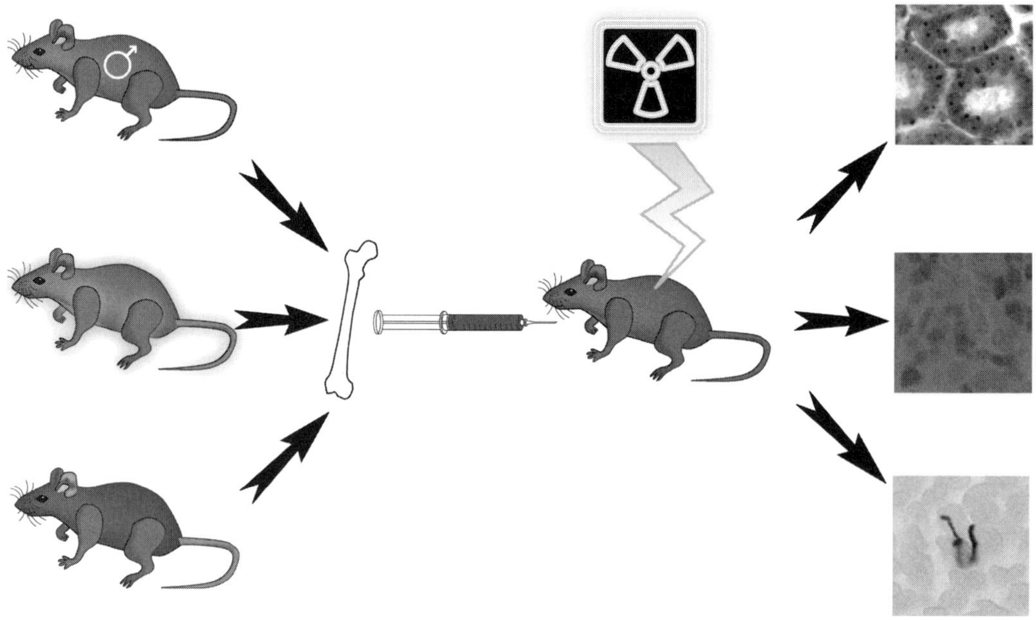

Fig. 12.1. Revealing that bone marrow cells (BMCs) have differentiated into non-haematopoietic cells can be achieved by transplanting lethally irradiated animals with new BMCs that can be tracked whatever their subsequent fate. This would include male BMCs to a female recipient, or GFP- or LacZ-positive BMCs to wild-type recipients. The male chromosome can be detected by in situ hybridisation, GFP by immunohistochemistry and β-galactosidase by X-gal histochemistry. (*see* Color Plate 11)

Human studies rely on the examination of tissue from patients that have either undergone a sex-mismatch liver transplant (usually female liver allografted to male recipient) or a sex-mismatch bone marrow transplant and then investigating biopsies for X and Y chromosome-expressing cells. In both humans and animals, these donor-derived cells can be assessed for phenotype using a variety of histochemical and immunohistochemical markers. A further modification is to use isotopically labelled RNA riboprobes to study the function of these donor-derived, phenotypically characterised cells. Taking advantage of the t9:22 that occurs in the majority of CML patients, resulting in the Philadelphia chromosome with the *BCR:ABL* fusion gene, we further show that bone marrow cell engraftment in the liver can occur without irradiation and bone marrow transplantation. (**Fig. 12.2C**)

We also demonstrate the detection of engrafted human cells of bone marrow origin in immunodeficient mice by distinguishing between human and murine cells using species-specific pan centromeric probes. (**Fig. 12.2B**)

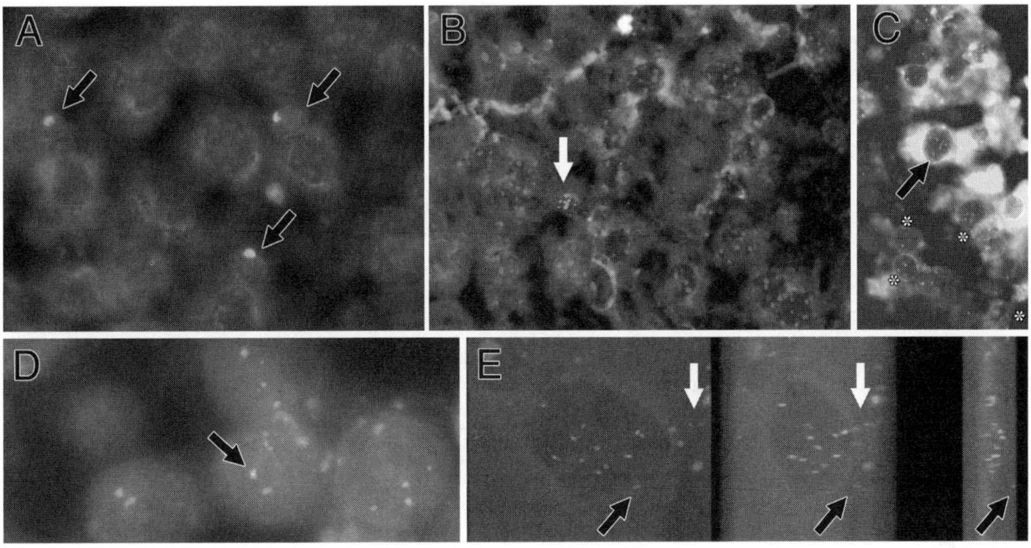

Fig. 12.2. Fluorescent and confocal microscopy. (**A**) Male cells (arrows) in male bone marrow-transplanted female mouse liver (green FITC dot). These cells are CK18 immunoreactive (red cytoplasm), suggestive of hepatocyte differentiation. (**B**) Human cell (green FITC, spotty nucleus, arrowed) in mouse liver (pink CY3 spots) after injection of human $CD133^{+}$ cells into a NOD-SCID mouse. (**C**) BCR/ABL probe on human liver in a case of CML showing normal ploidy, with two copies of chromosome 9 (red signals) and two copies of chromosome 22 (green signals) in some cells (asterisks), but multiple copies (polyploidy) in another cell (arrow). (**D**) BCR/ABL fusion signal (green and red overlap producing orange, arrowed) seen in cell tentatively identified as a hepatocyte in a case of CML. There is one native chromosome 9 (red), one native chromosome 22 (green) and one small red signal (*ASS* gene). (**E**) Confocal images demonstrating liver polyploidy in a female mouse transplanted with male bone marrow, with multiple X chromosomes (green signals) showing that a Y chromosome (red signal, black arrow) is outside the nuclear membrane (view **E**), while a smaller nucleus (white arrow) has both X and Y chromosomes contained within it. (*see* Color Plate 12)

2. Materials

2.1. Basic Histological Preparation of Sections

1. Neutral buffered formalin (BDH) tissue fixative: prepared as a 10% solution.
2. Coated slides (Fisher superfrost).
3. Graded alcohols 70, 96 and 100% prepared using Analar grade. ethanol and double-distilled water.
4. Xylene (BDH).
5. Wax (Lamb).

2.2. Pretreatments

1. Hydrogen peroxide (BDH) blocking solution for endogenous peroxidases. Add 2.4 ml 30% hydrogen peroxide to 100 ml absolute alcohol.
2. PBS tablets (Sigma): dissolve in double-distilled water according to the manufacturer's instruction.
3. Pepsin: dissolve 0.4 g pepsin (Sigma) in 0.1 M HCl (*see* **Note 1**)

4. Sodium thiocyanate (Sigma-Aldrich): dissolve 16 g in 200 ml of double-distilled water at 80°C.
5. Glycine stop: dissolve 0.4 g glycine (BDH) in double-strength PBS.
6. Paraformaldehyde (PFA) (Sigma): prepare a 4% solution by dissolving 4 g of powder in 100 ml PBS at 80°C. Cool to room temperature before use (*see* **Note 2**). Use on day of preparation.
7. Glass coverslips of various sizes to cover tissue section.
8. Rubber cement: available from cycle repair suppliers.
9. Humid chamber: this can be made from any sealable container large enough to take a slide rack horizontally – line with damp tissue soaked in distilled water.

2.3. Probes

1. Mouse Y chromosome paint (Cambio), FITC labelled.
2. Human X and Y chromosome paint (Stretton Scientific).
3. Mouse and human pan centromeric probes (Cambio).
4. BCR/ABL probe (Vysis).

2.4. Post-washes

1. Standard sodium citrate (SSC): to make the 20× stock, dissolve 175.3 g of sodium chloride and 88.2 g of sodium citrate in 900 ml water. Adjust to pH 7 with NaOH or HCl if necessary, make up to 1 l and sterilise by autoclaving. Dilute as necessary.
2. Anti-fluorescein POD (Roche): diluted 1:200 with PBS.
3. 3,3′-Diaminobenzidine (DAB): prepare a working solution by dissolving 6 mg in 10 ml of PBS, mix well and then add 20 μl 30% hydrogen peroxide.
4. Vectaset (Vector Labs) Hardset with DAPI.
5. Haematoxylin (Lamb).
6. DPX mountant (BDH).

2.5. Immunohis-to chemistry

1. Various primary antibodies raised against leukocyte common antigen, (CD45), cytokeratin 8/18, α-smooth muscle actin available from many sources including Dako, Novocastra, R&D Systems and Santa Cruz.
2. Biotinylated rabbit anti-mouse (Dako): use 1:300.
3. Biotinylated swine anti-rabbit (Dako): use 1:300.
4. Streptavidin-peroxidase (Dako): use 1:500.
5. Streptavidin–alkaline phosphatase (Dako): use 1:50.
6. Vector Red (Vector labs): prepare a working solution according to the manufacturer's instructions by adding two drops of stock to 5 ml of 100 mM Tris buffer pH 8.4.
7. Acid alcohol block: 20% acetic acid in absolute alcohol.

2.6. ISH for mRNA

1. Nuclease-free water (Q): add 0.1% vol of diethylpyrocarbonate (DEPC) to all solutions, then autoclave.
2. TE buffer: 10 mM Tris-HCl (pH 8.0), 1 mM EDTA.
3. RNA polymerases (Promega) T7, SP6 or T3 depending on the desired template.
4. RNAse inhibitor (Promega): use at a 20 U/μl concentration.
5. Dithiothreitol (DTT) (Sigma).
6. AGC mix (Boehringer): prepare as 6.25 mM aliquots of each individual base, ATP, GTP, CTP and then mix in equal volumes to give a final concentration of 1.0 mM.
7. ^{3}H UTP (GE Healthcare) ≈800 Ci/mmol.
8. DNaseI. RNase-free grade (Boehringer).
9. Chromaspin-30 columns DEPC-equilibrated (Clontech).
10. Transfer RNA (Sigma).
11. Triethananolamine buffer (Sigma): dissolve 37.5 g of triethanolamine in DEPC-treated water, then make up to 2 l to give a 0.1 M solution.
12. Acetic anhydride (Sigma): immediately before use, add 1.25 acetic anydride to 500 ml of 0.1 M triethanolamine (Acetylation buffer).
13. Formamide wash solution (Fisher): to prepare wash buffer add 250 ml of 10× salts (14.2 g Na_2HPO_4 in 300 ml Q pH 6.8, add 176.2 g NaCl), then add 100 ml of Tris-HCl pH 7.6 and 250 ml of 0.2 M EDTA, pH 7.5. Mix well and make up to 1 l. For 1 l of wash buffer solution, add 250 ml of 10× salts to 1.25 l of formamide, then make up to 2.5 l with Q.
14. Deionised Formamide (for use in hybridisation buffer): To 400 ml formamide add 20 g of ion exchange resin (20–50 mesh Bio Rad, Germany). Stir overnight then filter using No.1 filter paper to remove resin. Store at –20°C in small aliquots.
15. Dextran sulphate solution: dissolve 50 g of powder (Sigma) in 100 ml of autoclaved water at 80°C until dissolved. Aliquot in 1 ml tubes and store at –20°C.
16. Denhardt's salt solution: add 5 ml of 10× salts to a vial containing 5 ml Denhardt's solution and stir until dissolved. Transfer to a 50 ml tube and make up to 25 ml. Store in small volumes at –20°C.
17. Hybridisation buffer: for 1 ml, combine the following in order: 100 μl Denhardt's salt solution, 500 μl deionised formamide, 30 μl tRNA, 200 μl dextran sulphate (prewarmed), 10 μl 1 M DTTMix well. Store at –20°C
18. TNE buffer (Tris-HCl/NaCl/EDTA) for RNase digestion: combine 146 g of NaCl, 50 ml of 1 M Tris-HCl pH 7.6,

25 ml of 0.2 M EDTA, pH 7.5 and make up to 5 l with water. Adjust pH to 7.2–7.6.

19. RNaseA: dissolve 500 mg ribonucleaseA (Sigma) in 10 ml of 10 mM sodium acetate pH 5.2. Heat to 100°C for 15 min, cool, adjust pH to 7.4 with 1 M Tris-HCl.
20. Phenol:choloroform:isoamyl alcohol (PCI): prepared by mixing 24 ml of phenol and 24 ml chloroform, then adding 1 ml of isoamyl alcohol, before mixing well. The bottom phase is then used.
21. Chloroform:isoamyl alcohol (CI): prepared by adding 1 ml of isoamyl alcohol to 25 ml chloroform and using the bottom phase.

3. Methods

It is possible to use a variety of methods, and therefore markers, to investigate the origin and phenotype of cells. The simplest approach is to study the origin of cells using basic in situ hybridisation (ISH) and look for the Y chromosome either by direct fluorescence (**Fig. 12.2**) or by indirectly using light microscopy. (**Fig. 12.3**) Combining ISH with immunohistochemistry (IHC) further allows the investigator to determine the phenotype of the cells under investigation. These methods may also be combined with riboprobe ISH to look for an appropriate mRNA indicative of functionality, although this is technically very difficult and prone to failure. **Figure 12.3** illustrates the problems associated with combining these techniques.

3.1. Basic Histology

1. Tissue is harvested from experimental animals and fixed immediately in NBF. After fixing for a set period of time (*see* **Note 3**), the tissue is transferred to 70% alcohol and processed to wax blocks using standard histological techniques. Paraffin sections that are 4–6 μm thick are cut on a microtome, collected onto coated slides and dried overnight at 37°C.

3.2. ISH Pretreatments

1. Slides are dewaxed in xylene, 3 × 5 min.
2. Transferred to absolute alcohol and blocked in hydrogen peroxide if required (*see* **Note 4**) for 10 min before re-hydrating through graded alcohols to water and then PBS.
3. Wash in PBS for 15 min with three changes.
4. Treat slides in1 M sodium thiocyanate at 80°C for 10 min.
5. Wash in PBS, two changes over 10 min.
6. Digest slides in 0.4% pepsin at 37°C for required time (*see* **Note 5**).

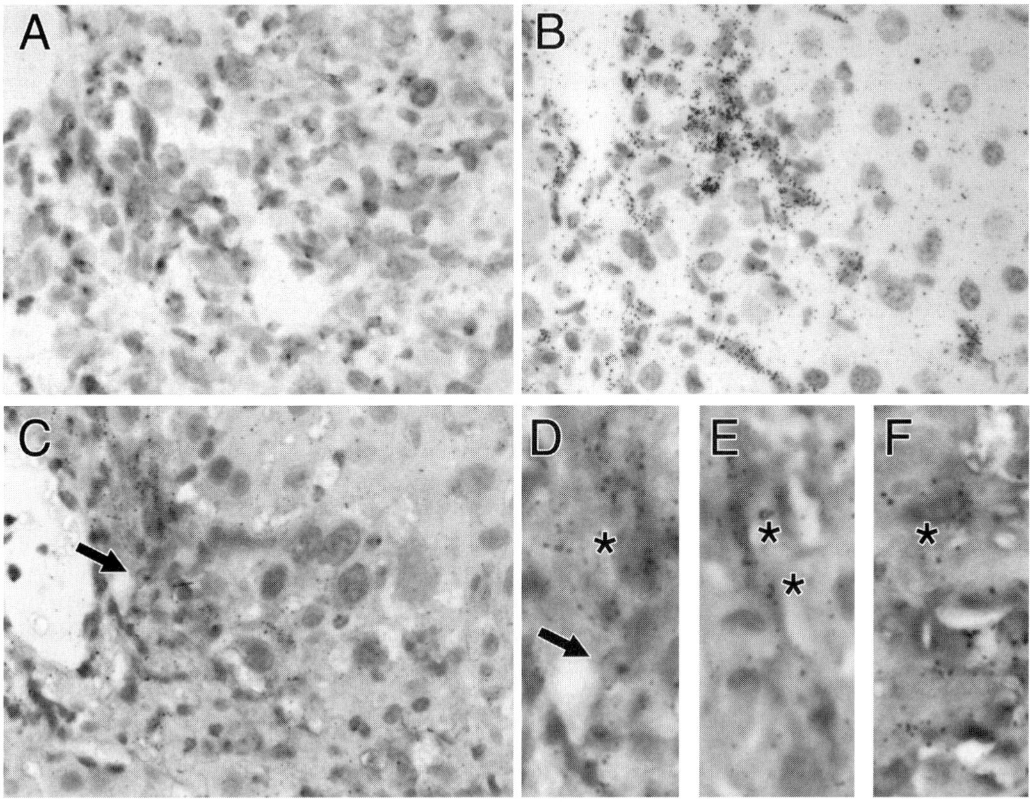

Fig. 12.3. Liver fibrosis in a mouse as viewed by bright field microscopy. (**A**) Demonstration of Y chromosome-positive cells (brown nuclear dots) in a female mouse liver after a male bone marrow transplant. (**B**) Demonstration of mRNA for pro(α1)I (black autoradiographic grains) in the same liver using a ^{3}H-labelled antisense riboprobe. (**C**) Demonstration of Y chromosome detection (brown dot, arrow) and IHC for α-SMA expression (red staining) – a marker of myofibroblast differentiation. (**D**) Demonstration of the expression of mRNA for pro(α1)I, the Y chromosome and α-SMA in the same liver. One Y chromosome-positive cell is expressing neither α-SMA nor mRNA for pro(α1)I, but another cell (asterisk) is expressing all three markers. Note the reduced grain density when techniques are combined in comparison to when ISH for the mRNA is performed alone. (**E** and **F**) Examples of ISH for pro(α1)I mRNA expression and immunoreactivity for α-SMA in the same section. (*see* Color Plate 13)

7. Stop the digestion by immersing in glycine stop for 5 min.
8. Wash in PBS for 5 min.
9. Post-fix in PFA for 2 min.
10. Wash well in PBS: three washes over 15 min.
11. Dehydrate through graded alcohols and air dry.
12. Remove probe from fridge or freezer and allow to warm to room temperature before applying 10–15 μl to each slide depending on the size of the tissue.
13. Cover with a glass coverslip and seal with rubber cement.
14. Denature the sealed slide at the required temperature for 10 min (*see* **Note 6**).
15. Place the slides horizontally in a humid chamber and hybridise overnight at 37°C.

3.3. ISH Post-hybridisation Washes and Detection

1. Carefully remove the rubber cement by rubbing gently between the thumb and the forefinger and then remove the coverslip.
2. Rinse the slides quickly in 0.5% SSC at 37°C, then wash for 5 min at 37°C, again in 0.5% SSC.
3. Wash with PBS, three changes over 15 min.
4. At this stage it is possible to mount the slides in Hardset and view using suitable software on a fluorescent microscope (*see* **Note 7**), see **Fig. 12.2** for examples.
5. Alternatively, sections that have been probed with a fluorescein-conjugated probe may be detected by applying 200 µl of anti-fluorescein POD FAB fragments to each slide for 1 h.
6. Wash with three changes of PBS over 15 min and detect using DAB for 2–3 min.
7. Wash with three changes of PBS over 15 min.
8. Counterstain with a light haematoxylin, dehydrate, clear in xylene and mount in DPX. In **Fig. 12.3**, panel A shows an example of a mouse Y chromosome paint on a section of female liver from a female mouse that had received a male bone marrow 6 weeks previously.

3.4. Immunohistochemistry

It is possible to combine ISH with IHC for a variety of specific cell markers. The IHC needs to be performed before ISH as the treatments involved in ISH destroy the IHC epitopes.

1. Dewax and block sections as for ISH. If using both peroxidase and alkaline phosphatase for detection, it will be necessary to also block for endogenous alkaline phosphatase for 5 min in ice-cold acid alcohol.
2. Take sections to PBS and wash for 5 min.
3. Perform any necessary antigen retrieval (*see* **Notes 8–10**).
4. Wash with three changes of PBS over 15 min.
5. Apply 1:25 blocking serum according to the species in which the secondary layer was raised for 15 min (*see* **Note 8**).
6. Apply primary antibody at a pre-determined dilution (*see* **Notes 9 and 10**) for 40 min in a humidity chamber, e.g. leukocyte common antigen (CD45) is diluted 1:200 if using a mouse monoclonal from Dako, but at 1:20 if using a rat monoclonal from Pharmacia.
7. Wash with three changes of PBS over 15 min.
8. Apply a second layer (biotinylated) for 40 min.
9. Wash with three changes of PBS over 15 min.
10. Apply a tertiary layer for 40 min. If continuing with FITC-labelled probe then Vector Red is the detection method of choice as it allows both direct and indirect visualisation;

therefore apply 1:200 streptavidin-AP. For DAB detection, apply 1:500 streptavidin-HRP.

11. Wash with three changes of PBS over 15 min.
12. Detect colour using either DAB or Vector Red (*see* **Note 10**) (see step 10 above).
13. Continue with ISH pre-treatments from step 4.

3.5. ISH Probe Preparation

The main technical problem with combining mRNA ISH for a functional marker such as the mRNA for the α1 chain of type I (pro)collagen [pro(α1) I] with dual DNA ISH and IHC is the preservation of the RNA during the necessary pre-treatments. Utmost care must be taken to avoid any source of contamination with ribonucleases. Riboprobes are labelled single-stranded RNA molecules synthesised by in vitro transcription using a DNA-directed RNA polymerase. The labelled UTP is usually ^{35}S but ^{3}H gives an improved spatial resolution due to its low-energy particles and is therefore the isotope of choice when deciding which cell is responsible for individual autoradiographic silver grains. As with IHC, the reader is advised to seek more expert help if adapting the basic methods in this way *(1)*.

1. After obtaining a plasmid containing the sequence of interest (usually in the form of an agar slope with bacteria transfected with the coding region in a suitable vector) this needs to be extracted using a kit such as Qiagen maxiprep and following the given instructions.
2. The plasmid is then linearised (to give template) using a suitable restriction endonuclease (*see* **Note 11**). For mRNA of pro(α1) I (IMAGE clone 335137), 200 units of *Eco*R1 will linearise 50 μg of purified plasmid, which can then be cleaned up using PCI.
3. Make a ^{3}H-labelled single-stranded mRNA pro(α1)I antisense probe using the DNA-directed RNA polymerase T3. Add to a microfuge tube at room temperature in the following order: 2.5 μl 5× transcription buffer (as supplied with polymerase), 1.0 μl RNase inhibitor, 0.7 μl DTT (100 mM), 2.0 μl AGC mix, 1.0 μg DNA template (made up to 2.4 μl with Q) 1 μl ^{3}H UTP, mix well then add 1.0 μl appropriate polymerase, in this case T3.
4. Incubate for 1 h at 37°C, then destroy the template by adding 1.0 μl DNase and incubate for a further 15 min. During this time prepare a Chromaspin 30 column by centrifugation according to the manufacturer's instructions.
5. Spin the tube to reduce the risk of radioactive aerosols, then add carrier RNA (10 mg/ml, 1.5 μl) with 10 mM DTT to a final volume of 25 μl (*see* **Note 12**).

6. Add bulk of reaction mix to top of the chromaspin column and centrifuge at 700 g for 3 min, collect the elute into a new tube containing 4 μl 100 mm DTT and 2 μl RNase inhibitor. Store at –20°C (*see* **Note 12**).

3.6. Radiolabelled ISH Pre- and Post-treatments

It is crucial to avoid RNase contamination. If combining with DNA ISH and /or IHC, then apply stringent precautions throughout all treatments. The steps must be undertaken wearing gloves, using sterile glassware and adding DEPC to all solutions. *(1, 2)*

1. Slides should be in PBS after any preceding treatments.
2. Permeabilise in proteinase K at 37°C for 10 min (*see* **Note 13**).
3. Rinse in glycine/2× PBS for 5 min to stop the protease.
4. Wash in two changes of PBS over 5 min.
5. Post-fix in PFA for 10 min.
6. Wash in PBS, three changes over 15 min.
7. Immerse slides in 500 ml acetylation buffer for 20 min (*see* **Note 14**).
8. Dehydrate through sequential alcohols and air dry. The slides are now ready to hybridise.
9. Calculate the volume of hybridisation mix required (*see* **Note 15**). Mix enough probe mix, allowing an extra 10% as it is difficult to pipette, mix well, then boil in a screw top tube for 2 min to denature before cooling on ice.
10. Apply 20 μl to each slide and cover with a glass coverslip. Place the slides horizontally in a suitable slide rack or slide mailing box, then place in a lunch box or similar container humidified with blotting paper soaked in 1× salts in 50% formamide (see **Note 16**) and place at 55°C overnight.
11. Pre-warm all solutions required the following day. A volume of 5 l of formamide wash buffer at 55°C, 5 l of TNE washes at 37°C, 1 l of 2× SSC and 500 ml of 0.5× SSC, both at 65°C.
12. The following day, remove the slides and gently ease off each coverslip by rubbing between the thumb and the forefinger. Place all slides in 500 ml formamide wash solution at 55°C on a rocking table. Keeping everything at 55°C, wash the slides with the full 5 l of wash buffer over the next 3–4 h.
13. Remove all traces of formamide by washing with the 4.5 l of TNE over 30 min. To the remaining 500 ml of TNE, add 1 ml of stock RNase.
14. Place the slides in RNAse solution at 37°C for 1 h.
15. Wash slides in 2× SSC at 65°C for 30 min twice.
16. Wash slides in 0.5× SSC at 65°C for 30 min.
17. Pass slides through graded alcohols and air dry.

3.7. Autoradiography (see Note 17)

1. In a darkroom fitted with a 902 filter and 15 W bulb, heat 25 ml water to 42°C in a cut-down measuring cylinder or similar in a suitable water bath. Check that the water bath does not have a light source indicating power or temperature.
2. Cool a metal plate.
3. Using plastic forceps or spoon to add strands of emulsion to the 40-ml mark, stir gently and leave to melt for at least 10 min, stirring occasionally.
4. Check that there are no bubbles in the solution by dipping a plain control slide, wipe the back and lay on a cooled metal plate to set. The dipping solution is ready to use if no bubbles can be seen on the test slide when it is held up to the light.
5. Dip slides one at a time, allow excess to drip off for a second, wipe the back and lay on a cooled plate to dry and set. This takes 1–2 h.
6. When dry, place all slides in a wooden or plastic box, seal in light-proof black bag and leave to expose at 4°C as appropriate (see **Note 18**).
7. Develop a set of exposed slides by immersing in a D19 developer for 4 min at 18°C for 4 min, stop in 1% acetic acid, wash in tap water and then fix in 30% sodium thiosulphate for 8 min. The main light may now be turned on.
8. Wash in running tap water for 1 h to remove any trace of fixative before counterstaining (*see* **Note 19**).

3.8. Discussion

The majority of the techniques described here are routinely used in many laboratories, but are rarely combined in the ways described. It is important to emphasise the absolute necessity to avoid contamination with any DNAses and or RNAses. Autoclaving of solutions and glassware, the wearing of gloves, the addition of DEPC to all solutions are all fundamental in preserving targets of interest *(1, 2)*.

The use of markers for the Y chromosome is widespread in the field of liver stem cell research *(3–8)* and in other organ systems *(9* and for review see *10)*; **Fig. 12.2A** illustrates the presence of cells of donor origin (male) after a sex-mismatch bone marrow transplant to a female mouse. **Figure 12.2B** illustrates the combination of two separate probes on a single section. NOD-SCID mice received an injection of $CD133^+$ haematopoietic stem cells 7 days prior to killing. The livers were examined using an FITC-labelled human pan centromeric probe combined with a CY3-labelled mouse pan centromeric probe. It is not possible here to determine whether the human cells are merely residing in the liver or whether they have undergone differentiation. This would necessitate combining these two probes with another technique such as IHC for phenotype or a riboprobe to show function. A complication of combining techniques is that the use of unmasking agents to allow for the detection of one signal is the very procedure that destroys a subsequent target. **Figure 12.3C** illustrates this point: signals for both RNA and Y chromosome are both greatly reduced when they are combined together.

The decision as to how to examine slides for the presence of the Y chromosome may depend on the availability of a suitable fluorescent microscope. The use of confocal microscopy further allows the examiner to determine the exact position of signals and confirm that they are inside or outside the nucleus – this is of particular importance in ploidy and translocation studies (**Fig. 12.2C–E**). The use of direct visualisation will also allow the investigator to determine true signals over background by looking in several channels but for researchers used to examining tissues by direct microscopy structural interpretation may be more difficult.

When attempting to use a triple method, it may be necessary to adapt the basic Y probe protocols to avoid high temperatures, which may destroy RNA. This may be done by denaturing the Y probe by boiling for 2 min before applying it to the slides and not co-denaturing at 60°C. As already mentioned, the combination of two techniques causes the loss of some signals, including a third method that is technically very challenging (**Fig. 12.3C–F**), but is possible provided that signals are very strong when performed individually (**Fig. 12.3B**).

4. Notes

1. Pepsin is known to autodigest. It is important to always allow it to dissolve at 37°C without too much agitation for a set time before use to standardise it. The powder is very light and should be handled with care.
2. PFA is hazardous. Wear gloves and avoid breathing vapour.
3. A standardised fixation time at this stage will avoid too greater variation in subsequent digestion times.
4. It is necessary to treat slides in hydrogen peroxide to block for endogenous peroxidase if using peroxidase detection later.
5. The time required for pepsin digestion is very variable. Advisable to try between 2 and 45 min to determine a suitable time when signal strength is adequate without losing too much morphology.
6. The denaturing temperature varies according to the probe used. Most mouse probes work well at 60°C, whereas human probes usually require 80°C.

7. We recommend Smartcapture software (Digital Scientific) although other similar software packages may be just as good. Depending on the fluorescent label used, DAPI, FITC and CY3 filters will be required, although using other filters will help to distinguish true signals from autofluorescence.
8. Basic immunohistochemistry is beyond the remit of this article. For those wishing to learn more about this subject, reference books such as the *Handbook of Immunochemical Staining Methods (11)* as supplied by Dako are recommended.
9. Many antibodies require antigen retrieval before the epitope can be revealed; techniques to achieve this include digestion in various proteases including trypsin, and heating by either boiling, microwaving or pressure cooking. It is necessary to refer to the data sheet supplied with individual antibodies to determine the appropriate method.
10. The use of appropriate controls cannot be over-emphasised. It is also usual to lose some signals due to the harshness of the ISH treatments, so it is recommended that colour is allowed to develop strongly before proceeding to ISH pretreatments.
11. Restriction endonucleases cleave DNA at known sites – the enzyme to use is determined by the orientation and position of the sequence of interest within the plasmid vector.
12. It is possible at his stage to take samples of pre- and post-spin column aliquots and count in a scintillation counter to assess the quantity of labelled probe made, and also to assess the quality by running on a 6% polyacrylamide denaturing gel (if using ^{35}S but not if using ^{3}H). Radiolabelled probes do not keep well. It is advisable to use within 3 days of preparation.
13. This step may not be necessary if tissue has already been digested previously.
14. This must be prepared immediately before use.
15. Add 20 µl per slide. Hybridisation buffer should make up 84% of the final volume. For ^{3}H-labelled probes, aim to add 200,000 counts to each slide, the difference in volume is made up with Q.
16. Formamide is toxic. Avoid breathing in fumes.
17. All autoradiography must be carried out under safe light conditions. Cleanliness is important, avoid the use of any metal coming into contact with the slides, i.e. use plastic racks.
18. This may be anything from 2 to 20 weeks depending on the strength of the signal.
19. Giemsa is the usual counterstain of choice if just looking at these slides; however, after DNA ISH and IHC, haematoxylin may be preferred as it gives a better contrast with vector

red and DAB. Silver grains indicating the presence of message appear as very small black dots under conventional light microscopy or as bright white dots when using dark field-reflected light illumination. It is usual to lose much of the signal when combining with other techniques.

Acknowledgements

We thank Prof. R. Revoltella for the tissue for demonstrating human cells in mouse liver.

References

1 Poulsom, R., Longcroft, J. M., Jeffery, R. E., et al. (1998) A robust method for isotopic riboprobe in situ hybridisation to localise mRNAs in routine pathology specimens. *Eur J Histochem* 42, 121–132.
2. Jeffery, R., Hunt, T., Poulsom, R. (2003) In situ hybridisation combined with immunohistochemistry to localise gene expression. Part IV Chapter 23, in (Brooks, S. A., Harris, A., eds.), *Breast Cancer Research Protocols*, pp. 323–346. Humana Press Inc.
3. Alison, M. R., Poulsom, R., Jeffery, R., et al. (2000) Hepatocytes from non-hepatic adult stem cells. *Nature* 406, 257.
4. Theise, N. D., Badve, S., Saxena, R., et al. (2000) Derivation of hepatocytes from bone marrow cells in mice after radiation induced myeloablation. *Hepatology* 31, 235–240.
5. Theise, N. D., Nimmakalu, M., Gardner, R., et al. (2000) Liver from bone marrow in humans. *Hepatology* 32,11–16.
6. Lagasse, E., Connors, H., Al-Dhalimy, M., (2000) Purified hematopoietic stem cells can differentiate into hepatocytes in vivo. *Nat Med* 6, 1229–1234.
7. Korbling, M., Katz, R. L., Khanna, A., et al. (2002) Hepatocytes and epithelial cells of donor origin in recipients of peripheral-blood stem cells. *N Engl J Med* 346, 738–746.
8. Sato, Y., Araki, H., Kato, J., et al. (2005) Human mesenchymal stem cells xenografted directly to rat liver are differentiated into human hepatocytes without fusion. *Blood* 106, 756–763.
9. Fang, T.-C., Alison, M. R., Cook, H. T., et al. (2005) Proliferation of bone marrow-derived cells contributes to regeneration after folic acid-induced acute tubular injury. *J Am Soc Nephrol* 16, 1723–1732.
10. Poulsom, R., Alison, M. R., Forbes, S. J., et al. (2002) Adult stem cell plasticity. *J Pathol* 197, 441–456.
11. Handbook of Histochemical Methods. 3rd Edition published by DAKO.

Chapter 13

Production of Hepatocyte-Like Cells from Human Amnion

Toshio Miki, Fabio Marongiu, Ewa C.S. Ellis, Ken Dorko, Keitaro Mitamura, Aarati Ranade, Roberto Gramignoli, Julio Davila and Stephen C. Strom

Abstract

Cells isolated from the placenta have been the subject of intense investigation because many of the cells express characteristics of multipotent or even pluripotent stem cells. Cells from the placental tissues such as amnion and chorion have been reported to display multilineage differentiation and surface marker and gene expression patterns consistent with embryonic stem (ES) and mesenchymal stem cells, respectively. We have reported that epithelial cells isolated from term placenta contain cells that express surface markers such as the stage-specific embryonic antigens (SSEA) and a gene expression profile that is similar to ES cells. When subjected to specific differentiation protocols, amniotic epithelial cells display markers of differentiation to cardiomyocytes, neurons, pancreatic cells and hepatocytes. If specific and efficient methods could be developed to induce differentiation of these cells to hepatocytes, the amnion may become a useful source of cells for hepatocyte transplants. Cells isolated from amnion also have some unique properties as compared to some other stem cell sources in that they are isolated from a tissue that is normally discarded following birth, they are quite plentiful and easily isolated and they do not produce tumors when transplanted. Cells isolated from the amnion may be a uniquely useful and noncontroversial stem cell source.

Key Words: Stem cell hepatocyte, hepatocyte transplant, cardiomyocyte, neuron, pancreatic beta cell, cell transplantation.

1. Introduction

While the transplantation of hepatocytes to treat liver disease has become a more common experimental technique worldwide, a major problem still exists concerning the source of cells for transplant *(1, 2)*. The most common source of cells has been from donor livers that have been rejected for transplantation because of steatosis, extended cold ischemic time, plaques in the vessels or moderate to advanced underlying liver disease *(3, 4)*. Thus, many

Anil Dhawan, Robin D. Hughes (eds.), *Hepatocyte Transplantation, vol. 481*

DOI 10.1007/978-1-59745-201-4_13 Springerprotocols.com

hepatocyte transplants rely on the isolation of cells for transplant from organs that were already judged not to be useful for transplant. Although it is clear that there are still useful cells in most organs that are not useful for whole-organ transplant, an alternative source of hepatocytes would increase the number of patients who could receive cell transplants. The most commonly proposed alternative sources of cells for hepatocyte transplants have been xenotransplants of hepatocytes from a porcine source, immortalized human cells, human fetal or progenitor cells or stem cell sources *(5)*. This chapter will discuss the possibility of generating hepatocytes from stem cells isolated from the amnion membrane of term placenta.

2. Amnion-Derived Stem Cells

Miki et al. *(6)* reported that epithelial cells from term human amnion (AE) have stem cell characteristics. Cell surface markers are commonly found on ES cells such as SSEA 3 and 4 and the tumor rejection antigens 1-60 and 1-81. In addition to the surface markers, AE cells also express molecular markers characteristic of ES cells including the expression of *Oct-4* and *Nanog*, genes known to be involved in the maintenance of pluripotency. The hypothesis that AE cells might be pluripotent was supported by the demonstration of the differentiation of the cells. Under certain culture conditions, AE cells differentiate to cell types derived from all three germ layers including cardiomyocytes, neurons, pancreatic alpha and beta cells and hepatocytes *(6–8)*. Work from other laboratories also supported the hypothesis that AE cells have stem cell characteristics *(9)*. Sakuragawa and co-workers *(10–13)* reported that AE cells could be induced to differentiate to cells with neural characteristics *(14, 15)*. These authors reported the expression of neural genes and proteins as well as the production and release of neurotransmitters *(16, 17)*. Wei et al. *(18)* explored the differentiation of AE cells to pancreatic cells and demonstrated the production and release of insulin and a lowering of blood glucose levels following the transplantation of AE cells into diabetic mice. Cells from the amniotic fluid and other compartments of the placenta also show stem cell characteristics, although most of the properties reported suggest that these cells are more similar to mesenchymal stem cells than AE cells *(19–26)*.

Taken together, the data clearly indicate that cells with stem or progenitor properties are located in the amnion membrane. If specific and efficient methods could be devised to induce differentiation of amnion-derived stem cells to hepatocytes, AE cells could be a useful source of stem cells for hepatocyte transplants. Amnion membrane is plentiful; with over 4 million live births per year in the United States, amnion-derived stem cells could be immediately available. Since relatively low-technology procures are needed for isolation and banking, this stem cells source could easily be available worldwide at a modest cost. Finally, this stem cell source is noncontroversial. Amnion membrane, like all other placental tissues, is normally discarded following the birth of a baby; thus tissue collection is not a problem. Since the life and development of the fetus is never interrupted, stem cells derived from placental tissues would be expected to avoid all of the ethical or religious concerns associated with some other stem cell sources.

3. Derivation of the Amnion and Isolation of AE Cells

Amnion develops during the second week of life (**Fig. 13.1**) at the time when the fertilized egg has begun implantation into the maternal endometrium and has formed a blastocyst *(7)*. While most of the fetal components of the placenta are derived from the hypoblast, the amnion differentiates from the epiblast, the same cell compartment that eventually gives rise to all organs and tissues of the developing fetus. The differentiation of amnion from the epiblast occurs before gastrulation and the specification of the three germ layers. Therefore, amnion might maintain some of the pluripotent nature of the epiblast. The differentiation of AE cells to cell types derived from all three germ layers supports this hypothesis. For the isolation of the

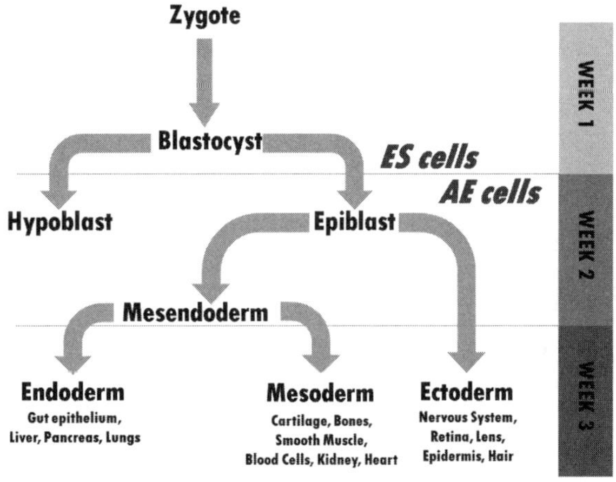

Fig. 13.1. Diagram of embryogenesis from fertilization to gastrulation.

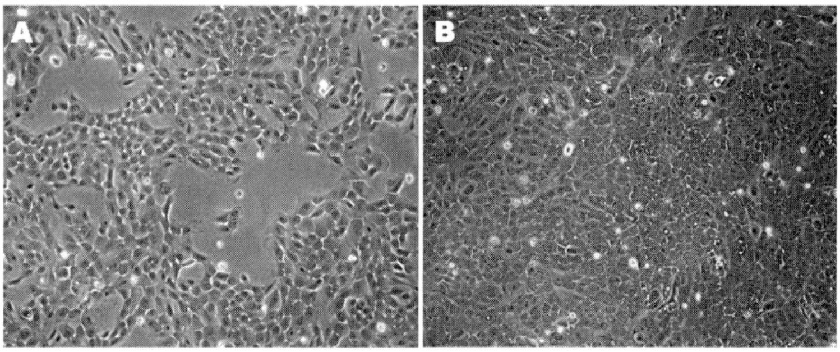

Fig. 13.2. AE cells in culture: low density (**A**) and high density (**B**).

stem cells, the amnion membrane is removed from the surface of the placenta. A detailed protocol has been published describing the separation of the amnion membrane from the placenta and the isolation of the epithelial cells *(27)*. Briefly, the amnion membrane is stripped from the surface of the placenta immediately following delivery. Amnion membrane is washed to remove blood and trypsinized to release epithelial cells. Trypsin digestions (one or two digestions of up to 40 min each) specifically release epithelial cells. Mesenchymal stromal cells (formerly called mesenchymal stem cells) remain in the amnion membrane through the procedure and, if needed, can be specifically released by digestion of the amnion membrane with collagenase following the removal of the epithelial cells. Following isolation, the epithelial cells can be immediately cryopreserved or placed in culture. Standard culture media consists of Dulbecco's Modified Eagle's Medium supplemented with 10% fetal bovine serum and 10 ng/ml epidermal growth factor (EGF), 1 mM nonessential amino acids (neaa), 4 mM L-glutamine (glu) 55 μM 2-mercaptoethanol (2ME). Specific growth factors can be added to help direct differentiation. Cultured cells grow to confluence quickly in the presence of serum and EGF. If EGF is removed, cell proliferation immediately slows, and then ceases even if serum supplementation is maintained. The morphology of human AE cells in culture at mid-confluence and complete confluence is presented in **Fig. 13.2**

4. Differentiation of AE to Hepatocyte-Like Cells

Differentiation of AE cells to different cell types is dependent on both the culture substrate and the types and concentration of growth factors added to culture media. Two simple protocols for

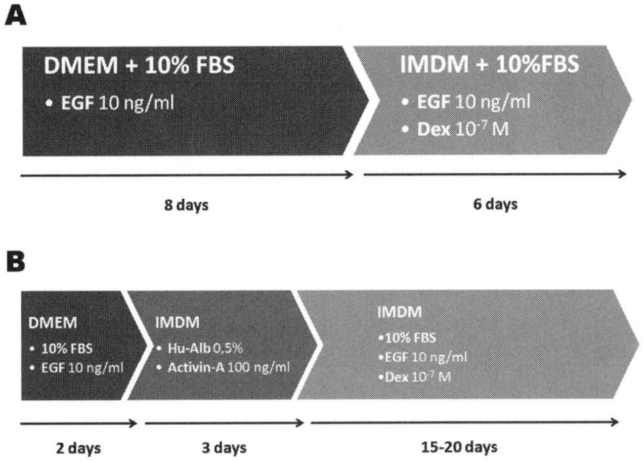

Fig. 13.3. Hepatocyte differentiation protocols.

hepatic differentiation are shown in the schematic form in **Fig. 13.3**. For hepatic differentiation, AE cells are plated on type 1 collagen-coated culture dishes in standard culture media supplemented with neaa, glu, 2ME and EGF as described above and steroid hormones such as hydrocortisone (HydC) or dexamethasone (Dex). Over the next days to weeks, the cells take on several markers of hepatic differentiation. We examined the ability of steroid hormone exposure to enhance hepatic differentiation. Data shown in **Fig. 13.4** show the relative expression of the endodermal/hepatic marker genes hepatocyte nuclear factor-4 (HNF4-α) and Alpha 1-antitrypsin (A1AT) at 7 and 14 days in culture in the presence of Dex or HydC. For these experiments, expression of the gene in the cells at the time of plating was set as 1 and the height of individual bars represent the relative expression of each gene at the indicated time points. Both HydC and Dex enhance the expression of the endodermal/hepatic genes in cultured AE cells. Since Dex was as good as or better than HydC in inducing endodermal/hepatic differentiation, the remaining studies were conducted with Dex as the steroid hormone in the media. Data presented in **Fig. 13.5** show the relative expression of albumin (Alb), A1AT and the liver-enriched

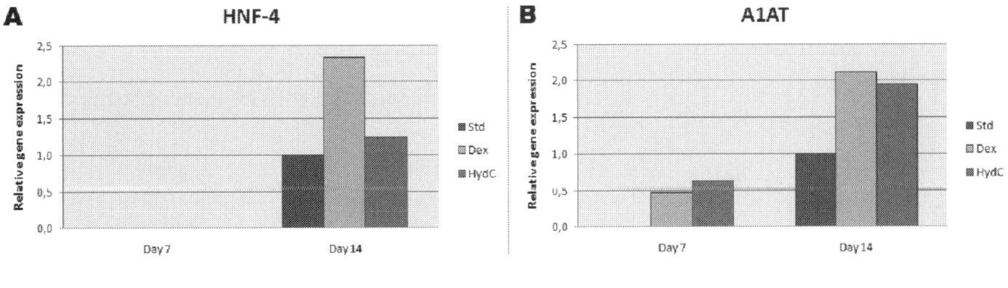

Fig. 13.4. Relative mRNA expression of HNF-4 (**A**) and A1AT (**B**) in AE cells cultured in the presence or absence of 1 μM Dex or HydC.

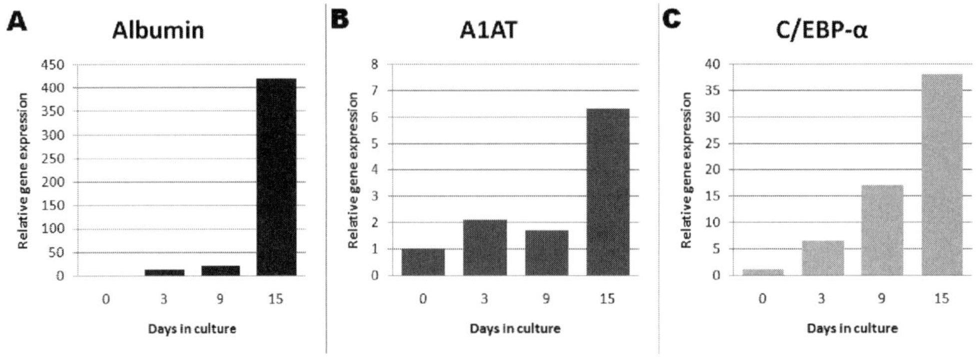

Fig. 13.5. Relative mRNA expression of albumin (**A**), A1AT (**B**) and C/EBP-α (**C**) in AE cells after 3, 9 and 15 days of culture in the presence of 0,1 μM Dex.

transcription factor, and CAAT enhancer binding protein-alpha (C/EBP-α). All three markers of hepatic differentiation increase over time in culture in the presence of Dex. Under these simple culture conditions Alb expression can increase over 400-fold and up to 35% of the cells will react positively to an antibody to human Alb.

Liver arises from the endoderm germ layer. The molecular events involved in endoderm formation have begun to be worked out from recent studies on zebra fish mutants, knockout mice and *Xenopus (28–31)*. These investigators described an intermediate stage of development between the mesoderm and the endoderm called the mesendoderm (**Fig. 13.1**) a bipotential tissue that gives rise to both mesoderm and endoderm. They also described growth factors and methods that enhance the formation of mesendoderm from undifferentiated cells. Mesendodermal differentiation of cells is accompanied by a decrease in the expression of stem cell marker genes and an increase in the expression of mesendodermal genes such as *FoxA2* and brachury (**Fig. 13.6**).

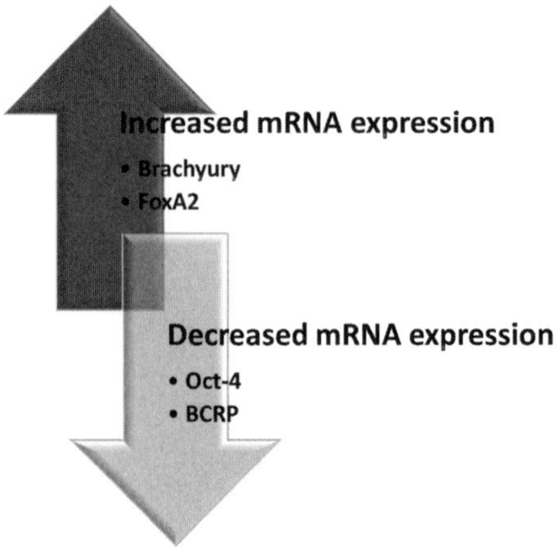

Fig. 13.6. Gene expression accompanying mesendodermal differentiation.

Differentiation of AE cells to an endodermal lineage such as liver might benefit from efficient differentiation of AE cells to mesendoderm, first, followed by exposure to additional growth factors that would enhance endodermal formation. One of the factors known to induce mesendodermal differentiation in undifferentiated cells is activin A, a member of the TGF-β superfamily. Activin A affects its biological activity via binding to the Type II, TGF-β cell surface receptor. Recent reports indicate that a brief exposure activin A enhances the differentiation of ES cells to the hepatic and pancreatic lineage *(32–35)*. However, other investigators reported that Activin A also plays an important role in maintaining self-renewal of ES cells *(36)*. Smith et al. *(37)* also showed that an inhibition of Activin signaling foster neurectoderm differentiation of ES cells.

During embryonic development, two types of endoderm are generated. Visceral endoderm contributes to the extraembryonic placental structures, while definitive endoderm gives rise to liver pancreas and other internal organs. Although many genes such as Alb and A1AT are expressed by both visceral and definitive endoderm, one gene that has been reported to be specific for definitive endoderm is CYP7A1 *(38)*. This gene is located on the endoplasmic reticulum and encodes cholesterol 7-alpha hydroxylase, an enzyme involved in the conversion of cholesterol to bile acids in hepatocytes *(39)*. The detection of expression of CYP7A1 in amnion-derived hepatocyte-like cells indicates that the AE cells differentiate to definitive endoderm. Other evidence of hepatic differentiation comes from studies of the regulation of CYP7A1 in AE cells. In the liver, the synthesis of bile acids from cholesterol is controlled in part by the regulation of the transcription of the CYP7A1 gene. The exposure of human hepatocytes in culture to bile acids results in a feedback inhibition of CYP7A1 expression *(40)*. We investigated the regulation of CYP7A1 expression in human AE-derived hepatocyte-like cells exposed to chenodeoxycholic acid (CDCA). The quantitative real-time reverse transcription-polymerase chain reaction (RT-PCR) data indicate that CYP7A1 expression is specifically downregulated by bile acid exposure **(Fig. 13.7)**. It is interesting that along with CYP7A1, bile salt export pump (BSEP) expression is also reduced by exposure to bile acids.

Data presented in **Table 13.1** show a partial list of the liver-specific or liver-enriched genes whose expression was detected in cultured AE cells exposed to differentiation conditions described in **Fig. 13.3**. Results shown in **Table 13.1** were generated by quantitative real-time RT-PCR or gene array studies. In addition to the expression CYP7A1, the expression of genes characteristic of *mature* human liver such as CYPs 1A2, 2B6 and 3A4 was detected. The wide range of hepatic genes detected in cultured AE, cells including the transcription factors, HNF4, C/EBP-alpha and beta, pregnane×receptor and constitutive androstane

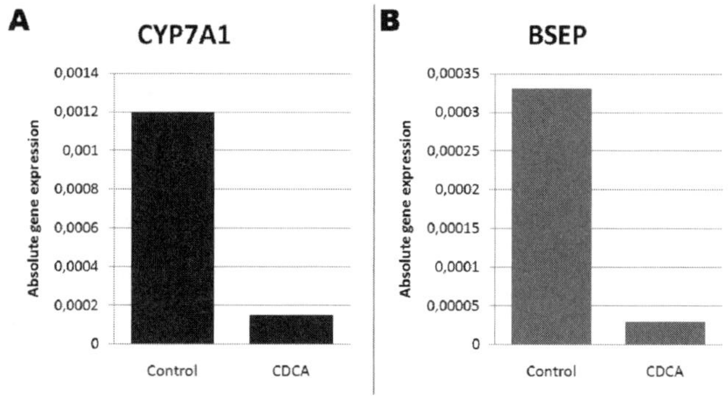

Fig. 13.7. Absolute mRNA expression of CYP7A1 (**A**) and BSEP (**B**) in AE cells cultured in the presence or absence of CDCA.

receptor, all of the other CYP genes and genes encoding hepatic transport proteins (**Table 13.1**) provide additional evidence that AE cells follow a pathway to definitive endoderm and authentic hepatic differentiation. From the initial studies with the differentiation protocols provided here, the level of expression of the mature liver genes such as the CYP enzymes range from approximately 0.5 to 16% of the values normally expressed in mature human liver. We noticed that the levels of expression of the individual genes are similar to those observed in human fetal liver at mid-gestation (18–22 weeks). Some genes are expressed preferentially during the fetal period and decline in expression following birth. A well-known example of this type of pattern can be found with alpha fetoprotein (AFP) and Alb. Fetal liver expresses high levels of AFP, but expression of this gene declines rapidly after birth. Alb expression increases throughout gestation and remains high during adult life. Genes in the CYP3A family show a similar pattern to AFP and alb. Fetal liver expresses predominantly CYP3A7, while mature liver expresses predominantly

Table 13.1
In vitro differentiation of hAE cells to hepatocyte-like cells

• Cytokeratines: 8, 18, 19
• Albumin, Alpha 1-antitrypsin, C-met
• CYP7A1
• HNF-1, HNF-4α, C/EBPα, C/EBPβ, OATP, PXR, CAR, RAR, RXR and PPAR…
• CYP450 gene expressions: *1A2, 2B6, 2C8, 2C9, 2C19, 2D6, 3A4, 3A7* and *7A1*

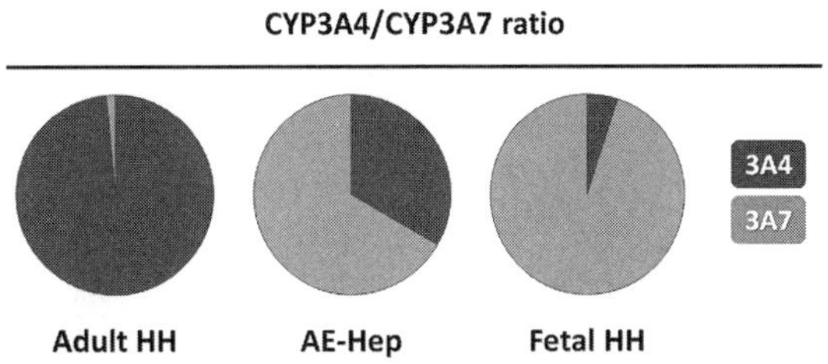

Fig. 13.8. The ratio of the expression of CYP3A4 to 3A7 RNA in adult and fetal liver and in cultured AE cells induced to hepatic differentiation.

CYP3A4 *(41–43)*. This is presented in a schematic form in (**Fig. 13.8**). In cultured AE hepatocyte-like cells, CYP3A7 comprises approximately 60% of the total CYP3A gene expression. The expression of both CYP3A7 and CYP3A4 suggests that AE differentiates along a pathway similar to authentic fetal human liver. The relative ratio of 3A4 to 3A7 suggests that cells are progressing toward mature hepatocytes.

Other laboratories have reported similar observations of hepatic differentiation of AE cells *(44)*. Longer and perhaps more complex differentiation and/or selection protocols will be needed before cells with a full adult liver phenotype are produced in vitro. Optimization of differentiation protocols requires the investigation of multiple growth factors in dose–response type experiments. The number of possible combinations becomes quite large. We have employed a high-throughput, microscope-enabled instrument to aid in the optimization of hepatic differentiation protocols. Some select growth factors were screened for their potential to induce the nuclear translocation of HNF4-α with an ArrayScan. The ArrayScan VTI (Cellomics, Pittsburgh, PA, USA) is an automated fluorescence microscopic with integrated image analysis and data management systems that allows high-content screening analysis on cultured cells. Naive AE cells (20×10^3 cells per well) were plated on a 96-well plate. In the experiments summarized in **Fig. 13.9,** cells were cultured 1 week with the growth factors, EGF, FGF-8, FGF-19, hepatocyte growth factor (HGF), OSM, DEX at concentrations from 6.25 to 400 ng/ml. Control cultures were not exposed to growth factors. Immunofluorescence analysis was performed with a rabbit anti-human HNF-4α antibody and the corresponding Cy3-conjugated secondary antibody. The fluorescent intensity of the nucleus was measured and an equivalent size area of cytoplasm was measured and quantified. From these data, a nuclear/cytoplasmic ratio can be calculated to identify growth factor treatments that enhance nuclear localization of HNF4-α. Cell images (1,300/well) were acquired. Results presented in **Fig. 13.9** show that EGF is a potent inducer of the nuclear localization of HNF4-α and that

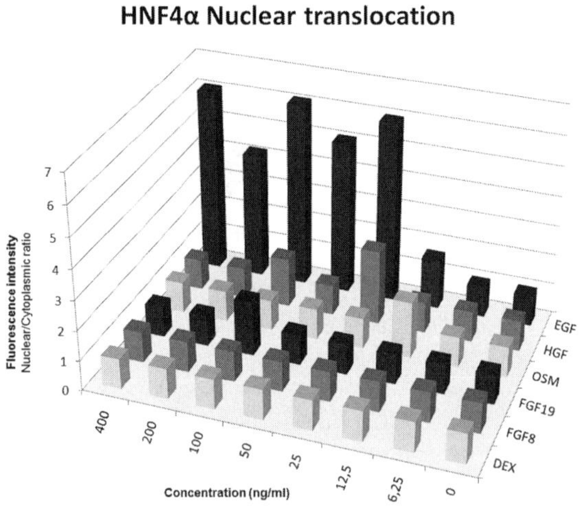

Fig. 13.9. ArrayScan analysis was performed to investigate the effect of multiple growth factors and the various concentrations of each growth factor. The localization of HNF-4α protein was indicated by the nuclear/cytoplasmic ratio of fluorescent intensity.

HGF also has a similar but smaller effect. These multiwell image analysis techniques allow the rapid analysis of multiple growth factors and multiple concentrations of each in small-volume assays. Such studies should shorten the time required to optimize differentiation protocols. In addition to in vitro studies, the transplantation of undifferentiated or partially differentiated AE cells into the liver of suitable recipients may provide a microenvironment, which more completely supports and instructs the hepatic differentiation of AE cells. Such studies are under way.

5. Evidence of Bipotential Hepatic Differentiation of AE

The coexpression of AFP and cytokeratin 19 in AE cells is reminiscent of a stage in hepatic development during fetal life where the liver is inhabited by bipotential progenitor cells, which can give rise to both hepatocytes and bile ducts. The coexpression of these two genes in AE cells might suggest that they are bipotential as well, and that in addition to hepatocytes, the generation of bile ducts might also be possible. Attempts were made to induce differentiation of AE to biliary cells. Published reports indicate that endothelial cells will undergo tube formation when plated on a substrate of matrigel followed by exposures to growth factors such as vascular endothelial growth factor (VEGF). AE cells were

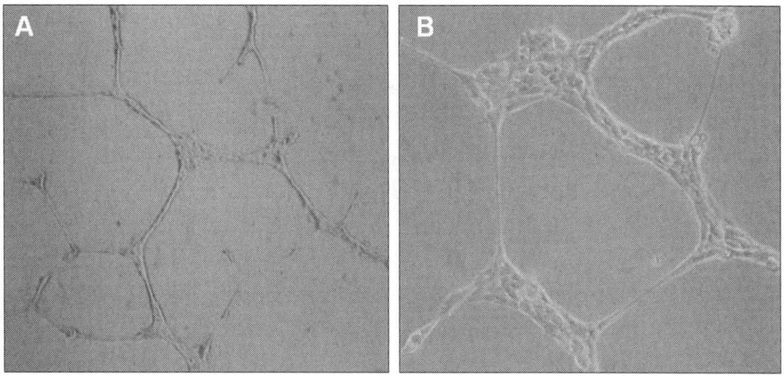

Fig. 13.10. AE cells form web-like ductular structures on Matrigel™-coated culture plates: ****A = 40x; B = ****100x.

plated at low seeding density on matrigel-coated plates under standard growth conditions, which included EGF but specifically did not include VEGF. The results of these experiments are shown in **Fig. 13.10** Within 4 days, AE cells arranged themselves in clusters of cells connected by long tubes of cells. In these experiments, web formation was dependent on the presence of the matrigel substrate as these three-dimensional structures did not form on plastic or collagen-coated culture plates. Subsequent analysis indicated that the tubes contain a rudimentary lumen and the cells react strongly with antibodies to cytokeratin 19 (data not shown). The expression of the epithelial marker, CK19, clearly indicates that the duct-like structures are not endothelial, vascular structures. The observations of ductular formations of epithelial cells that express CK19 are consistent with the differentiation of AE cells to biliary or pancreatic ducts. Additional research is under way to determine if more mature epithelial ducts with a large and complete lumen can be produced from AE, and if in addition to a physical similarity, the epithelial ducts express genes, proteins and functions common to biliary or pancreatic ducts. If so, these ductular structures might be useful for reconstruction of extrahepatic biliary duct defects or the regeneration of other ductular structures.

6. Conclusions

Data presented here and in previous reports clearly indicate that human term amnion contains cells with characteristics commonly found in pluripotent stem cells such as ES cells including the expression of surface markers and genes that maintain pluripotency. In addition, AE cells display the ability to differentiate into cell types from all three germ layers, in vitro. In this report, we present some initial studies to identify useful protocols to induce mesendodermal and subsequent endodermal differentiation of AE cells to cells with hepatic characteristics. If effective and efficient procedures can be developed to induce hepatic differentiation and purify hepatocyte-like cells from AE cultures, these cells may become a useful and noncontroversial cell type for hepatocyte transplantation and regenerative medicine. If hepatocyte transplant procedures are to be employed on large numbers of patients, plentiful sources of hepatocytes for transplantation will be needed. Stem cell sources such as AE hold the promise of providing these much-needed cells.

Acknowledgements

This research was supported in part by a grant from Pfizer, Inc.

References

1. Strom, S., Bruzzone, P., Cai, H., et al. (2006) Hepatocyte transplantation: clinical experience and potential for future use. *Cell Transpl* 15, S105–S10.
2. Fisher, R. A., Strom, S. C. (2006) Human hepatocyte transplantation: worldwide results. *Transplantation* 82, 441–9.
3. Nakazawa, F., Cai, H., Miki, T., et al. (2002) in *Proceedings of Falk Symposium, Hepatocyte Transplantation*, Vol. 126, pp. 147–58, Kouwer Academic Publishers, Lancaster, UK.
4. Mitry, R. R., Hughes, R. D., Dhawan, A. (2002) Progress in human hepatocytes: isolation, culture & cryopreservation. *Semin Cell Dev Biol* 13, 463–7.
5. Strom, S., Fisher, R. (2003) Hepatocyte transplantation: new possibilities for therapy. *Gastroenterology* 124, 568–71.
6. Miki, T., Lehmann, T., Cai, H., et al. (2005) Stem cell characteristics of amniotic epithelial cells. *Stem Cells* 23, 1549–59.
7. Miki, T., Strom, S. C. (2006) Amnion-derived pluripotent/multipotent stem cells. *Stem cell reviews* 2, 133–42.
8. Davila, J. C., Cezar, G. G., Thiede, M., et al. (2004) Use and application of stem cells in toxicology. *Toxicol Sci* 79, 214–23.
9. Elwan, M. A., Sakuragawa, N. (1997) Evidence for synthesis and release of catecholamines by human amniotic epithelial cells. *Neuroreport* 8, 3435–8.
10. Elwan, M. A., Ishii, T., Ono, F., et al. (1999) Evidence for the presence of dopamine D1 receptor mRNA and binding sites in monkey amniotic epithelial cells. *Neurosci Lett* 262, 9–12.
11. Elwan, M. A., Ishii, T., Sakuragawa, N. (2003) Characterization of the dopamine transporter gene expression and binding sites in cultured human amniotic epithelial cells. *Neurosci Lett* 342, 61–4.
12. Sakuragawa, N., Thangavel, R., Mizuguchi, M., et al. (1996) Expression of markers for both neuronal and glial cells in human amniotic epithelial cells. *Neurosci Lett* 209, 9–12.
13. Sakuragawa, N., Misawa, H., Ohsugi, K., et al. (1997) Evidence for active acetylcholine metabolism in human amniotic epithelial cells: applicable to intracerebral allografting for neurologic disease. *Neurosci Lett* 232, 53–6.
14. Tamagawa, T., Ishiwata, I., Saito, S. (2004) Establishment and characterization of a pluripotent stem cell line derived from human amniotic membranes and initiation of germ layers in vitro. *Hum Cell* 17, 125–30.
15. Ilancheran, S., Michalska, A., Peh, G., et al. (2007) Stem cells derived from human fetal

membranes display multi-lineage differentiation potential. *Biol Reprod* 77, 577–588.

16. Kakishita, K., Elwan, M. A., Nakao, N., et al. (2000) Human amniotic epithelial cells produce dopamine and survive after implantation into the striatum of a rat model of Parkinson's disease: a potential source of donor for transplantation therapy. *Exp Neurol* 165, 27–34.
17. Kakishita, K., Nakao, N., Sakuragawa, N., et al. (2003) Implantation of human amniotic epithelial cells prevents the degeneration of nigral dopamine neurons in rats with 6-hydroxydopamine lesions. *Brain Res* 980, 48–56.
18. Wei, J. P., Zhang, T. S., Kawa, S., et al. (2003) Human amnion-isolated cells normalize blood glucose in streptozotocin-induced diabetic mice. *Cell Transplant* 12, 545–52.
19. In 't Anker, P. S., Scherjon, S. A., Kleijburg-van der Keur, C., et al. (2004) Isolation of mesenchymal stem cells of fetal or maternal origin from human placenta. *Stem Cells* 22, 1338–45.
20. In 't Anker, P. S., Scherjon, S. A., Kleijburg-van der Keur, C., et al. (2003) Amniotic fluid as a novel source of mesenchymal stem cells for therapeutic transplantation. *Blood* 102, 1548–9.
21. Prusa, A. R., Marton, E., Rosner, M., et al. (2003) Oct-4-expressing cells in human amniotic fluid: a new source for stem cell research? *Hum Reprod* 18, 1489–93.
22. Tsai, M. S., Lee, J. L., Chang, Y. J., et al. (2004) Isolation of human multipotent mesenchymal stem cells from second-trimester amniotic fluid using a novel two-stage culture protocol. *Hum Reprod* 19, 1450–6.
23. Tsai, M. S., Hwang, S. M., Tsai, Y. L., et al. (2006) Clonal amniotic fluid-derived stem cells express characteristics of both mesenchymal and neural stem cells. *Biol Reprod* 74, 545–51.
24. De Coppi, P., Bartsch, G., Jr., Siddiqui, M. M., et al. (2007) Isolation of amniotic stem cell lines with potential for therapy. *Nat Biotechnol* 25, 100–6
25. De Coppi, P., Callegari, A., Chiavegato, A., et al. (2007) Amniotic fluid and bone marrow derived mesenchymal stem cells can be converted to smooth muscle cells in the cryo-injured rat bladder and prevent compensatory hypertrophy of surviving smooth muscle cells. *J Urol* 177, 369–76.
26. Parolini, O., Alviano, F., Bagnara, G. P., et al. (2007) CONCISE REVIEW: Isolation and Characterization of Cells from Human Term Placenta: Outcome of the First International Workshop on Placenta Derived Stem Cells. *Stem Cells.*
27. Miki, T., Marongiu, F., Ellis, E., et al. (2007) Isolation of amniotic epithelial cells. *Current Protocols in Stem Cell Biology* 1E.3, 1E.3.1–1E.3.9.
28. Alexander, J., Stainier, D. Y. (1999) A molecular pathway leading to endoderm formation in zebrafish. *Curr Biol* 9, 1147–57.
29. Zoltewicz, J. S., Gerhart, J. C. (1997) The Spemann organizer of Xenopus is patterned along its anteroposterior axis at the earliest gastrula stage. *Dev Biol* 192, 482–91.
30. Woo, K., Fraser, S. E. (1997) Specification of the zebrafish nervous system by nonaxial signals. *Science* 277, 254–7.
31. Darras, S., Marikawa, Y., Elinson, R. P., et al. (1997) Animal and vegetal pole cells of early Xenopus embryos respond differently to maternal dorsal determinants: implications for the patterning of the organiser. *Development* 124, 4275–86.
32. D'Amour, K. A., Agulnick, A. D., Eliazer, S., et al. (2005) Efficient differentiation of human embryonic stem cells to definitive endoderm. *Nat Biotechnol* 23, 1534–41.
33. D'Amour, K. A., Bang, A. G., Eliazer, S., et al. (2006) Production of pancreatic hormone-expressing endocrine cells from human embryonic stem cells. *Nat Biotechnol* 24, 1392–401.
34. McLean, A. B., D'Amour, K. A., Jones, K. L., et al. (2007) Activin a efficiently specifies definitive endoderm from human embryonic stem cells only when phosphatidylinositol 3-kinase signaling is suppressed. *Stem Cells* 25, 29–38.
35. Soto-Gutierrez, A., Kobayashi, N., Rivas-Carrillo, J. D., et al. (2006) Reversal of mouse hepatic failure using an implanted liver-assist device containing ES cell-derived hepatocytes. *Nat Biotechnol* 24, 1412–9.
36. Xiao, L., Yuan, X., Sharkis, S. J. (2006) Activin A maintains self-renewal and regulates fibroblast growth factor, Wnt, and bone morphogenic protein pathways in human embryonic stem cells. *Stem Cells* 24, 1476–86.

37. Smith, J. R., Vallier, L., Lupo, G., et al. (2008) Inhibition of Activin/Nodal signaling promotes specification of human embryonic stem cells into neuroectoderm. *Dev Biol* 313, 107–117.
38. Asahina, K., Fujimori, H., Shimizu-Saito, K., et al. (2004) Expression of the liver-specific gene Cyp7a1 reveals hepatic differentiation in embryoid bodies derived from mouse embryonic stem cells. *Genes Cells* 9, 1297–308.
39. Ellis, E., Goodwin, B., Abrahamsson, A., et al. (1998) Bile acid synthesis in primary cultures of rat and human hepatocytes. *Hepatology* 27, 615–20.
40. Ellis, E. C. S. (2003) in *Department of Medicine*, Karolinska Institute, Stockholm.
41. Yang, H. Y., Lee, Q. P., Rettie, A. E., et al. (1994) Functional cytochrome P4503A isoforms in human embryonic tissues: expression during organogenesis. *Mol Pharmacol* 46, 922–8.
42. Schuetz, J. D., Beach, D. L., Guzelian, P. S. (1994) Selective expression of cytochrome P450 CYP3A mRNAs in embryonic and adult human liver. *Pharmacogenetics* 4, 11–20.
43. Wrighton, S. A., Molowa, D. T., Guzelian, P. S. (1988) Identification of a cytochrome P-450 in human fetal liver related to glucocorticoid-inducible cytochrome P-450HLp in the adult. *Biochem Pharmacol* 37, 3053–5.
44. Takashima, S., Ise, H., Zhao, P., et al. (2004) Human amniotic epithelial cells possess hepatocyte-like characteristics and functions. *Cell Struct Funct* 29, 73–84.

Chapter 14

Generation of Hepatocytes from Human Embryonic Stem Cells

Niloufar Safinia and Stephen L Minger

Abstract

Use of human hepatocytes for therapeutic and drug discovery applications is hampered by limited tissue source and the inability of hepatocytes to proliferate and maintain function long-term in vitro. Human embryonic stem (hES) cells are immortal and pluripotent and may provide a cell source for functional human hepatocytes *(1)* Here we have outlined some of the protocols currently in use for the generation of hepatocytes from hES cells.

Key words: Human embryonic stem (hES) cells and hepatocytes.

1. Introduction

Although liver transplantation has become an accepted treatment for acute and end-stage liver disease, the scarcity of organ donors limits its potential. Transplantation of hepatocytes has, therefore, been proposed as an aid and an alternative to whole-organ transplantation *(2)*. Hepatocytes, derived from unused livers, have been transplanted into the liver or ectopic sites such as the spleen and have been shown to support liver function. Although clinically used, deriving hepatocytes by this means is also limited, meaning other sources for hepatocytes need to be found to enable wider use of the treatment.

One possibility to derive hepatocytes for transplantation is the use of embryonic stem cells. In the last 20 years, mouse ES cells have served as a major biological tool for studying early embryonic development *(3)*. These pluripotent cells, isolated from the blastocyst-stage embryos, have been shown to differentiate into derivatives of the three embryonic germ layers. During in vitro differentiation, mouse ES cells have also been shown to develop into specialized somatic cells, including hepatocytes *(4, 5)*. The isolation of human

Anil Dhawan, Robin D. Hughes (eds.), *Hepatocyte Transplantation, vol. 481*

DOI 10.1007/978-1-59745-201-4_14 Springerprotocols.com

ES cells several years ago, however, expanded the potential of ES cells as a source of cells not only for developmental studies but also for cell therapy *(6)*. The pluripotency of human embryonic stem (hES) cells has been proven both in vivo and in vitro. In vivo studies have shown that injection of hES cells into immune-deficient mice can lead to the generation of teratomas, harboring all three embryonic germ layers *(7)*. In vitro studies in which hES cells are aggregated in suspension cultures leading to the formation of embryoid bodies (EBs) have been shown to express molecular markers specific to the three embryonic germ layers *(8)*. Here, we have summarized some of the protocols currently in use for the differentiation of hES along a hepatocyte lineage. In general, the methods of differentiation are divided into spontaneous and directed differentiation with the latter being subdivided into two categories: addition of growth factors and hormones and constitutive expression of hepatic transcription factors *(3)*.

1.1. Generation of hES Cells and 'Directed' Differentiation

As mentioned in the previous section, hES cells are pluripotent cells derived from the inner cell mass of in vitro-fertilized human preimplantation blastocysts. A study by Schuldiner et al. *(9)* in 2000 examined the potential of eight different growth factors including hepatocyte growth factor (HGF) and nerve growth factor (NGF) to direct the differentiation of hES cells in vitro. They showed that hES cells that had initiated development as EBs express a receptor for each of the factors. Differentiation of cells along a hepatocyte lineage was assessed by expression of cell-specific molecular markers.

1.2. Directed Differentiation of hES Cells Using Genetic Selection

Directed differentiation of hES cells into hepatic-like cells was first demonstrated by Lavon et al. in 2004 *(7)*. Using DNA microarray analysis they identified several genes expressed at high levels in either fetal or adult liver cells. Most of these genes were also shown to be expressed in hES cells undergoing differentiation as EBs, and in order to further explore the hepatic differentiation pattern of hES cells, they genetically engineered hES cells by placing the green fluorescent protein (GFP) reporter gene under the control of the albumin promoter *(7)*. The cells were then sorted from the heterogeneous population of differentiating human ES cells using fluorescent activated cell sorting (FACs), and the population of hepatic-like cells was expanded through the addition of various growth factors. The protocol is outlined below.

1.3. Directed Differentiation of hES Cell Using Sequential Growth Factor Stimulation

An alternative means of directing hepatic differentiation is to sequentially induce differentiation first to definitive endoderm followed by directed differentiation into hepatic cells. Cai et al. *(10)* first used Activin A for 3 days to specifically direct undifferentiated hES cells toward an endodermal cell fate. At this time, more than 80% of cells expressed immunoreactivity for antigens indicative of definitive endoderm. Targeted differentiation of definitive endoderm to a hepatic lineage was achieved by the combined addition of the growth factors, FGF4 and BMP2, for 5 days. These primitive hepatic cells were further differentiated to mature functional hepatocytes by the sequential addition of HGF, oncostatin M and dexamethasone. These cells were shown to have properties of functional human hepatocytes including the ability to secrete albumin, to store glycogen and to take up indocyanine green. These cells were also shown to be capable of engraftment in the spleen and migration into the liver of immune-compromised CCl_4-treated mice.

2. Materials

2.1. Cell Culture

1. Human ES cells [H9 clone *(11)*] were grown on mouse embryo fibroblasts in 80% knockout DMEM, an optimized Dulbecco's modified Eagle's medium for ES cells (Gibco/BRL), 20% knockout SR, a serum-free formulation (Gibco/BRL), 1 mM glutamine (Gibco/BRL), 0.1 mM β-mercaptoethanol (Sigma), 1% nonessential amino acids stock (Gibco/BRL), 4 ng/ml basic fibroblast growth factor (bFGF) (Gibco/BRL) and 10^3 units/ml leukaemia inhibitory factor (LIF) (Gibco/BRL).
2. 0.1% gelatin (MERCK) was used to cover tissue culture plates.
3. To induce formation of EBs, 10^7 hES cells were transferred by using 0.1%/1 mM trypsin/EDTA (Gibco/BRL) to 100 mm^2 low-adherence plastic petri dishes to allow their aggregation and to prevent adherence to the plate.
4. Human EBs were grown in the same culture medium, except in the absence of LIF and bFGF.
5. EBs were cultured for 5 days after which time they were dissociated with trypsin and plated on 100 mm^2 tissue culture plate coated with 50 μg/ml fibronectin (Boehringer Mannheim).
6. Cells were grown in the presence of various factors including 20 ng/ml hepatocyte growth factor (HGF) (R&D Systems, Minneapolis, MN, USA) and 100 ng/ml β-nerve growth factor (β-NGF) (R&D Systems). Of note, all examined growth factors were absent from the commercially available knockout serum replacement in which the embryonic stem cells were cultured (**Fig. 14.1**).

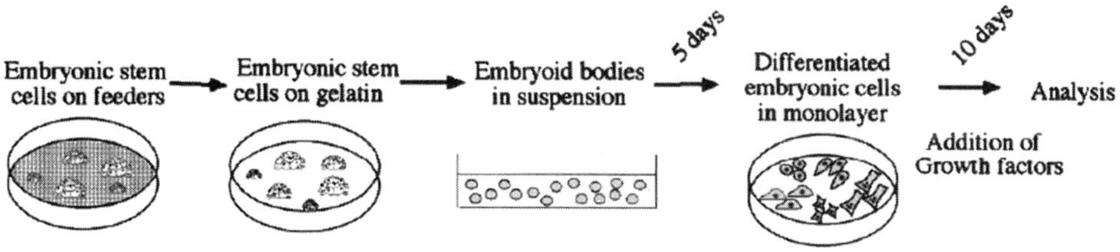

Fig. 14.1. A schematic representation of the differentiation protocol (taken from ref. *(9)* with permission).

2.2. Reverse Transcription-Polymerase Chain Reaction Analysis

1. Total RNA was extracted by using an Atlas Pure Total RNA labeling Kit (Clontech, Franklin Lakes, NJ, USA).
2. cDNA was synthesized from 1 μg total RNA, by using an Advantage RT-for-PCR Kit (Clontech).
3. cDNA samples were subjected to PCR amplification with DNA primers selective for human gene sequences.
4. For each gene, the DNA primers were derived from different exons to ensure that the PCR product represents the specific mRNA species and not genomic DNA.
5. PCR was performed by using the Clontech AdvanTaq plus RT-PCR kit and by using a two-step cycle at 68°C.
6. Primers were synthesized for the following human genes (specific to hepatocytes): albumin, α1-anti-trypsin and α-feto protein.

2.3. Analysis

1. Initially the presence of growth factor receptors was determined at the stage when growth factors were to be added to the culture (**Fig. 14.2**).
2. RNA from human ES cells, 5-day-old EBs and 10-day-old differentiated hES cells (DE) was isolated and analyzed by reverse transcription-PCR (RT-PCR) by using primers specific to receptors for the various growth factors.
3. The differentiation of hES cells induced by growth factors was further examined by determining the expression of cell-specific genes by using RT-PCR (**Fig. 14.3** and *see* **Note 1**).

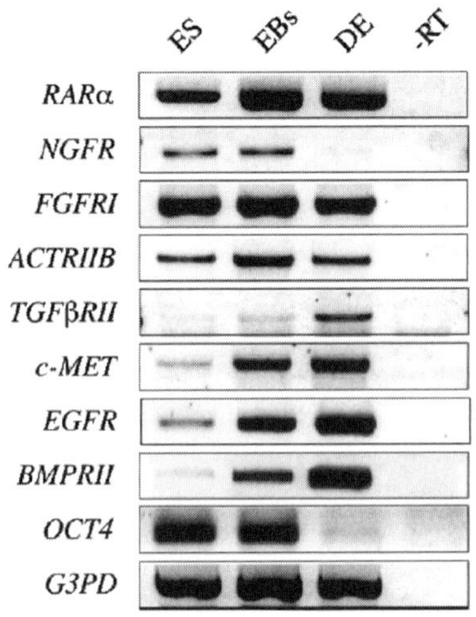

Fig. 14.2. Expression of receptors for various growth factors in human embryonic cells. RNA samples from ES cells, 5-day-old EBs and 10-day-old DE cells were analyzed by RT-PCR for expression of specific receptors (taken from ref. *(9)* with permission).

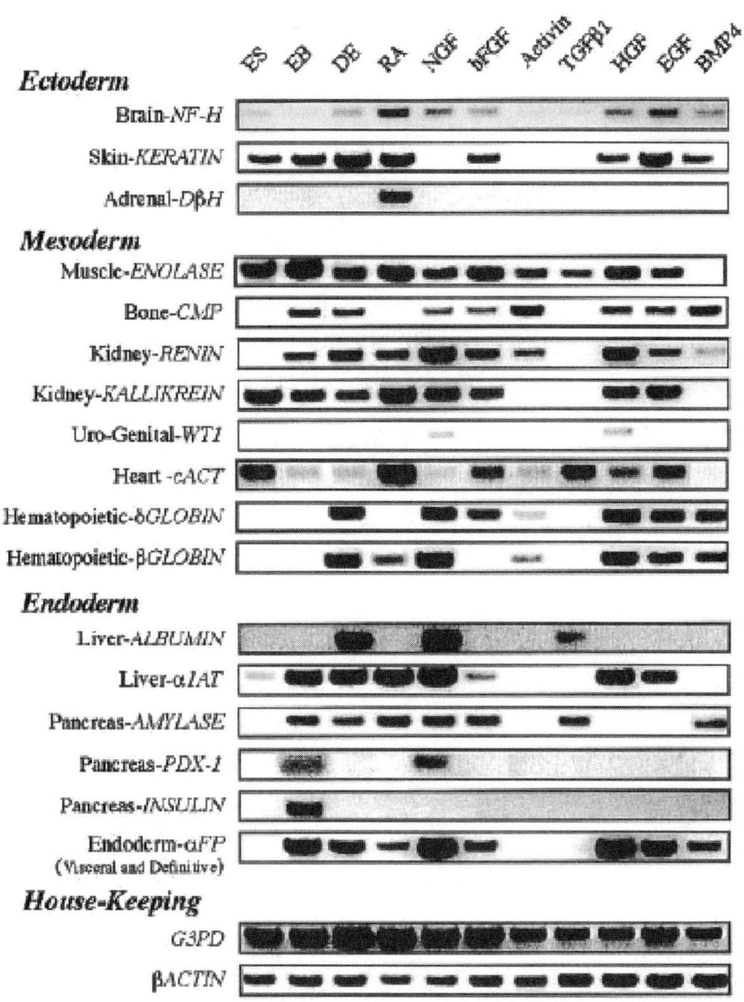

Fig. 14.3. Analysis of expression of cell-specific genes in human ES cells treated with various growth factors. RNA from ES, 20-day-old EBs and DE cells treated with different growth factors was analyzed by RT-PCR for expression of cell-specific genes and two house-keeping genes (taken from ref. *(9)* with permission).

3. Methods

3.1. Cell Culture

1. Human ES cells and their differentiated derivatives were cultured as previously described either as EBs or as differentiated ES cells *(9)*.
2. For teratoma formation, 5×10^6 hES cells were injected into the testis of 4-week-old severe combined immunodeficiency (SCID) mice.
3. After 1 month the mice were killed and the teratoma removed and frozen in liquid nitrogen. (All animal experiments were performed according to NIH guidelines.)

3.2. RT-PCR Analysis

1. Total RNA was extracted and 1 μg of RNA was reverse transcribed by random hexamer priming using an EZ-first Strand cDNA Synthesis Kit (Biological Industries, Kibbutz Beit Haemek, Israel).
2. cDNA samples were subjected to PCR amplification with DNA primers specific to the human genes using a pair of oligomers, each from a different exon.
3. All RT-PCR experiments were performed under non-saturation conditions. (PCR conditions include a first step of 3 min at 94°C, a second step of 20–30 cycles for 30 s at 94°C, a 30-s annealing step at 60–62 and 45 s at 72°C, and a final step of 5 min at 72°C
4. A description of primers and size of final products is described in **Table 14.1**.
5. Final products were assessed by gel electrophoresis on 2% agarose ethidium bromide-stained gels and their identity was verified through direct sequencing.

3.3. Plasmid Construction

1. The ALB-eGFP expression vector was constructed by deleting the CMV promoter sequence from peGFP-N1 (Clontech *(12)* and inserting the mouse albumin minimal promoter sequence into the *Hin* dIII restriction site.

Table 14.1
Primers used for PCR and size of final products

Gene	5' primer	3' primer	Size
APOA4	GTGGCAAGAAACTCCTCCAG	CCTTCCCAATCTCCTCCTTC	353 bp
APOB	ACCCGGAGAAAGATGAACCT	GAAGAGGTGTTGCTCCTTGC	371 bp
APOH	GCACTGAGGAAGGAAAATGG	GGCCATCCAGAGAATATCCA	357 bp
APOF	GGAAGCGATCAAACCTACCA	ATCAGCCTGACAACCAGCTT	347 bp
FGA	TCTCATCACCCTGGGATAGC	AAAAGCCATCCTCCCAAACT	338 bp
FGB	GGGAGAAAACAGGACCATGA	ATTGGGGACTATTGATGTCC	312 bp
FGG	GAATTTTGGCTGGGAAATGA	TGTTCAGCACAGTTGCCTTC	314 bp
AFP	AGAACCTGTCACAAGCTGTG	GACAGCAAGCTGAGGATGTC	676 bp
ALB	GTGAGACCAGAGGTTGATGTG	CATTCATGAGGATCTGCAGCG	760 bp
ADHIC	TGCAGGAATCTGTCGTTCAG	GAAGGTGCTGACGCCGAC	312 bp
GAPDH	AGCCACATCGCTCAGACACC	GTACTCAGCGCCAGCATCG	302 bp

2. The construct contained an SV40-driven neomycin selectable marker, which confers resistance to G418 antibiotic.
3. Transfection and establishment of cell lines were performed as previously described *(12)*.

3.4. Fluorescence-Activated Cell Sorter Analysis and Cell Sorting

1. Analysis was performed on a FACSCalibur system (Becton Dickinson, Franklin Lakes, NJ, USA) according to green fluorescent emission for detection of enhanced GFP-positive cells.
2. Analysis was performed by CELLQUEST software (Becton Dickinson).
3. Forward- and side-scatter plots were used to exclude dead cells and debris from the histogram analysis plots.
4. The sorting of eGFP-positive cells was performed as described previously *(12)*.

3.5. Immunostaining

1. DE cells were washed several times and grown overnight with serum-free media in order to avoid cross-reaction with serum proteins.
2. The cells were then washed three times with saline and fixed onto the plate with 4% paraformaldehyde.
3. Either rabbit anti-human α-fetoprotein (Dako, Carpinteria, CA, USA) or monoclonal mouse anti-human albumin (Fitzgerald, Concord, MA, USA) was used as primary antibodies.
4. Secondary antibodies used included Cy-3-conjugated donkey anti-rabbit IgG (H+L; Jackson ImmunoResearch, West Grove, PA, USA) or CY-3-conjugated goat anti-mouse IgG (H+L; Jackson ImmunoResearch).
5. Teratomas were embedded in the OCT compound (Sakura Finetek USA Inc., Torrance, CA, USA) and 6 μm sections were stained using either anti-human albumin (Fitzgerald) or anti-α-cardiac actin (Maine Biotechnology Services Inc., Portland, ME, USA) antibodies.

3.6. Cytokine Treatments

1. Twenty-four-day-old EBs were dissociated and plated as DE cells for an additional 10 days with or without growth factors.
2. Growth factors were added as follows: 100 ng/ml acidic fibroblast growth factor (as previously described **ref.** *(13)*; Boehringer-Mannheim GmbH), 5 ng/ml bFGF (**ref** *(13)*; Boehringer-Mannheim GmbH), 20 ng/ml HGF (**ref.** *(14)* ; R&D Systems) and 50 ng/ml bone morphogenic protein 4 (**ref.** *(15)* ; R&D Systems). Additionally, conditioned medium from cultured mouse hepatocytes was also used to influence hepatic differentiation (*see* **Note 2**).

3.7. Cell Culture

1. Human ES cells H1 and H9 were used in this study and were propagated on irradiated mouse embryonic fibroblasts in DMEM/F12 medium containing 20% serum replacement,

1 mM glutamine, 0.1 mM β-mercaptoethanol, 1% nonessential amino acids (all from Gibco/Invitrogen) and 4 ng/ml bFGF (Peprotech).

2. To induce endodermal differentiation, hES cells were propagated in 1640 medium (Hyclone) supplemented with 0.5 mg/ml albumin fraction V (Sigma) and 100 ng/ml Activin A for 1 day. On day 2, 0.1% of insulin-transferrin-selenium (ITS, Sigma) was added to this medium and on the third day this was increased to 1% ITS.
3. To induce differentiation to a hepatic lineage, cells were then cultured in hepatocyte culture medium (Cambrex) supplemented with 30 ng/ml FGF4 (Peprotech) and 20 ng/ml BMP2 (Peprotech) for five days, with the media changed every day. To demonstrate that both of these factors were required for hepatic differentation, 20 ng/ml Su5402 (FGF4-receptor antagonist, Chemicon) or 800 ng/ml Noggin (BMP inhibitor, R&D Systems) was used in combination with FGF4 and BMP2.
4. To further differentiate hepatic cells to functional hepatocytes, cells were then propagated in 20 ng/ml HGF (Peprotech) for 5 days and then in 10 ng/ml Oncostatin M (R&D Systems) and 0.1 μm dexamethasone (Sigma) from then onwards.

3.8. RT-PCR Analysis

1. Total RNA was extracted by using TRIzol reagent (Invitrogen).
2. cDNA was synthesized from total RNA, by using the reverse transcription kit (Promega).
3. cDNA samples were subjected to PCR amplification with DNA primers (**Table 14.2**).
4. RT-PCR was performed by using the EXTaq polymerase and the following programe conditions: first step of 5 min at 94°C, 35 cycles for 30 s at 94°C, a 30-s annealing step at 50–57°C and 30 s at 72°C, and extension for 10 min at 72°C.

3.9. Real-Time RT-PCR

1. Real-time PCR was performed on an ABI Prism 7300 Sequence Detection System.
2. Reaction conditions consisted of 12.5 μl SYBR Green PCR Master Mix (ABI), 0.8 μl 10 mm forward and reverse primers, 10.4 μl water and 0.5 μl template cDNA in 25 μl reaction volume.
3. Reaction conditions were programed for 2 min at 50°C, 10 min at 95°C, followed by 40 cycles of 15 s at 95°C and 1 min at 60°C. Relative expression levels were normalized against the β-actin gene.

3.10. Immunofluorescence

1. Cells were fixed with 4% paraformaldehyde in phosphate-buffered saline (PBS) at room temperature for 20 min.
2. Non-specific antibody binding was inhibited by incubating the cells in 0.1% Triton X-100, 10% horse serum and 1% bovine serum albumin at room temperature for 1 h.

Table 14.2
Primers and Conditions Used for RT-PCR

Gene name	Primer sequence	Product length (bp)	Annealing temperature (°C)
AFP	Sense: TTTTGGGACCCGAACTTTCC Antisense: CTCCTGGTATCCTTTAGCAACTCT	451	56
Alb	Sense: GGTGTTGATTGCCTTTGCTC Antisense: CCCTTCATCCCGAAGTTCAT	502	56
CK8	Sense: GGAGGCATCACCGCAGTAC Antisense: TCAGCCCTTCCAGGCGAGAC	472	56
CK18	Sense: GGTCTGGCAGGAATGGGAGG Antisense: GGCAATCTGGGCTTGTAGGC	460	56
G6P	Sense: GCTGGAGTCCTGTCAGGCATTGC Antisense: TAGAGCTGAGGCGGAATGGGAG	350	56
AAT	Sense: ACATTTACCCAAACTGTCCATT Antisense: GCTTCAGTCCCTTTCTCGTC	183	56
HNF4α	Sense: CCACGGGCAAACACTACGG Antisense: GGCAGGCTGCTGTCCTCAT	290	56
PEPCK	Sense: CTTCGGCAGCGGCTATGGT Antisense: TGGCGTTGGGATTGGTGG	383	50
TDO2	Sense: TACAGAGCACTTCAGGGAG Antisense: CTTCGGTATCCAGTGTCG	285	50
TAT	Sense: CCCCTGTGGGTCAGTGTT Antisense: GTGCGACATAGGATGCTTTT	345	56
Cyp7A1	Sense: GTGCCAATCCTCTTGAGTTCC	397	57

Table 14.2
(continued)

Gene name	Primer sequence	Product length (bp)	Annealing temperature (°C)
	Antisense: ACTCGGTAGCAGAAAGAATACATC		
Cyp3A4	Sense: ATGAAAGAAAGTCGCCTCG Antisense: TGGTGCCTTATTGGGTAA	267	56
Cyp2B6	Sense: AGGGAGATTGAACAGGTGATT Antisense: GATTGAAGGCGTCTGGTTT	253	56
GAPDH	Sense: AATCCCATCACCATCTTCC Antisense: CATCACGCCACAGTTTCC	382	56

3. Cells were incubated in primary antibody overnight at 4°C.
4. Cells were washed five times with PBS and FITC- or TRITC-conjugated secondary antibodies (1:150 dilution, Santa Cruz) were applied for 1 h at 37°C.
5. 1 mg/ml DAPI (Roche) was used to counterstain the nuclei of each cell.
6. Primary antibodies included Sox 17 (R&D Systems), HFN3β (Upstate), CK-7, CK-18, CK-19, AAT (all from Invitrogen), α-fetoprotein and albumin (both from Dako).

3.11. Albumin Secretion

The concentration of albumin secreted into the tissue culture medium was quantified by ELISA using a Human Albumin ELISA Quantitation Kit (Bethyl Labs) with the values normalized to total cellular protein content (Micro BCA Protein Assay Kit, Pierce).

3.12. Transplantation of hES-Derived Hepatic Cells

1. Ten-week-old female SCID mice were used in this study. All procedures were approved by the Peking University Institutional Animal Care and Use Committee.
2. One day prior to transplantation, each animal was administered 10 ml of CCl_4 diluted 1:10 in sterilized mineral oil.
3. Approximately 100 μl of DMEM containing one million hepatic cells differentiated for 18 days were injected into the spleen of recipient animals. Control animals received an equal volume of DMEM alone. Animals were killed 8 weeks post-implantation, the livers were isolated and embedded in OCT compound and 7-mm-thick microtome sections were obtained for processing.

4. Transplanted human cells were identified by positive immunoreactivity for anti-human nuclear antigen (Chemicon) and anti-human AAT protein (Invitrogen).

4. Notes

1. It is apparent from **Fig. 14.3** that genes representative of liver cells are already expressed in EBs and differentiated hES cells in the absence of growth factors. The addition of NGF would appear to stimulate this expression pattern, although no quantitative or functional data were presented to verify this. However, this study is informative in demonstrating that hES cells spontaneously differentiate into cells that express markers of liver cells.
2. This protocol establishes conditions for the selection of highly enriched populations of hepatic cells using gene transfer of liver-specific promoter sequences and cell sorting technology. Conditioned medium from mouse liver hepatocytes was shown to significantly enhance hepatic differentiation of EBs derived from hES cells suggestive of paracrine factors that require further investigation.
3. This protocol follows one of the first successful reports of the significant enrichment of functional hepatic cells from hES cells using developmental cues to direct differentiation.

References

1. Rambhatla, L., Chiu, C-P., Kundu, P., et al. (2003) Generation of hepatocyte-like cells from human embryonic stem cells. *Cell Transplant* 12, 1–11.
2. Horslen, S. P., Fox, I. J. (2004) Hepatocyte transplantation. *Transplantation* 77, 1418–1486.
3. Lavon, N., Benvenisty, N. (2005) Study of hepatocyte differentiation using embryonic stem cells. *J. Cell. Biochem.* 96, 1193–1202.
4. Jones, E. A., Tosh, D., Wilson, D. I., et al. (2002) Hepatic differentiation of murine embryonic stem cells. *Exp Cell Res* 272, 15–22.
5. Kania, G., Blyszczuk, P., Jochheim, A., et al. (2004) Generation of glycogen and albumin producing hepatocyte-like cells from embryonic stem cells. *Biol Chem* 385, 943–953.
6. Reubinoff, B. E., Pera, M. F., Fong, C. Y., et al., (2000) Embryonic stem cell lines from human blastocysts: somatic differentiation in vitro. *Nat Biotechnol* 18, 399–404.
7. Lavon, N., Yanuka, O., Benvenisty, N. (2004) Differentiation and isolation of hepatic-like cells from human embryonic stem cells. *Differentiation* 72, 230–238.
8. Itskovitz-Eldor, J., Schuldiner, M., Karsenti, D., et al. (2000) Differentiation of human embryonic stem cells into embryoid bodies comprising the three embryonic germ layers. *Mol Med* 6, 88–95.
9. Schuldiner, M., Yanuka, O., Itskovitz-Eldor, J., et al. (2000) Effects of eight growth factors on the differentiation of cells derived from human embryonic stem cells. *PNAS* 97 (21): 11307–11312.

10. Cai, J., Zhao, Y., Liu, Y., et al. (2007) Directed differentiation of human embryonic stem cells into functional hepatic cells. *Hepatology* 45, 1229–1239.
11. Thomson, J. A., Itskovitz-Eldor, J., Shapiro, S. S., et al. (1998) *Science* 282, 1145–1147.
12. Eiges, R., Schuldiner, M., Drukker, M., et al. (2001) Establishment of human embryonic stem cell-transfected clones carrying a marker for undifferentiated cells. *Curr Biol* 11, 514–518
13. Jung, J., Zheng, M., Goldfarb, M., et al. (1999) Initiation of mammalian liver development from endoderm by fibroblast growth factor. *Science* 284, 1998–2003.
14. Schmidt, C., Bladt, F., Goedecke, S., et al. (1995) Scatter factor/ hepatocyte growth factor is essential for liver development. *Nature* 373, 699–702.
15. Rossi, J. M., Dunn, N. R., Hogan, B. L. et al. (2001) Distinct mesodermal signals, including BMPs from the septum transversum mesenchyme, are required in combination for hepatogenesis from the endoderm. *Genes Dev* 15, 1998–2009.

Chapter 15

Isolation, In Vitro Cultivation and Characterisation of Foetal Liver Cells

Yue Wu, Chetan C. Shatapathy, and Stephen L. Minger

Abstract

Hepatocyte transplantation has recently become an efficient clinical method in the treatment of patients with metabolic liver diseases. The shortage of donor cells remains an obstacle to treat more patients. Foetal liver tissues may therefore be useful as an alternative source of generating functional hepatocytes after in vitro culture and maturation.

Key words: Foetal liver tissue, in vitro culture, clonal selection.

1. Introduction

Liver failure is a serious health problem in the world. Currently, liver transplantation is the ultimate therapy for people suffering from end-stage liver diseases and liver-based metabolic diseases, but the scarcity of donor organs and the risk of surgical complications involved further complicate this approach. In recent years, hepatocyte transplantation has become an attractive therapeutic alternative to liver transplantation *(1–4)*.

The liver has the remarkable ability to massively regenerate following injury until it almost regains its normal mass, and hepatocytes are capable of rapid proliferation and functional regain following surgical resection *(5)*. However, hepatocytes are notoriously difficult to maintain and expand in vitro. The problems include difficulties in identifying, isolating and culturing pure cell populations, the limited proliferation capacity of cultured cells and their inability to retain a hepatic phenotype beyond a few passages *(6)*. Mature hepatocytes, in particular, show little potential to grow in primary culture *(7, 8)*, and even under

Anil Dhawan, Robin D.Hughes (eds.), *Hepatocyte Transplantation, vol. 481*

DOI 10.1007/978-1-59745-201-4_15 Springerprotocols.com

the influence of primary mitogens they enter into a limited number of divisions before de-differentiating with a rapid loss of tissue-specific genes *(9)*, or degeneration *(10)*. Besides, the paucity of molecular markers and their variability depending on culture conditions and cell preparation make identification and quantification of adult liver stem cells difficult. Complex culture techniques are involved in the isolation and culture of adult mouse liver progenitor cells and results in low yields of proliferating epithelial cells that can generate hepatocyte-like cells *(11)*.

Given the problems involved in the isolation, culture and maintenance of adult liver cells in vitro, culture of foetal liver tissue presents a good alternative experimental model to analyse mechanisms of hepatic development and regeneration. Compared to adult progenitor cells, foetal progenitor cells are comparatively uncommitted and are at an early developmental stage. Moreover, their pre-immune character, higher telomerase activity and resistance to cryopreservation make them attractive candidates for investigating potential clinical applications.

Various protocols have been described for the isolation of liver progenitor cells from newborn rodents *(12–18)* and from adult rat liver *(19)*. Recently, protocols to isolate multipotent progenitor cells from human foetal liver have also been demonstrated *(20, 21)*. Here we describe protocols for isolating hepatic progenitor cells from both embryonic day 14 rats and human foetal liver tissues, preferentially expanding and extensively replicating them in vitro in defined culture media, and generating pure clonal populations of cells derived from a single cell. The progenitor cells can be characterised by immunofluorescence, differentiated into mature hepatocytes using various differentiation protocols and function of the mature hepatocytes confirmed by albumin assay.

2. Materials

2.1. Isolation and Culture of Foetal Liver Cells

1. Fischer rat 344 embryos, embryonic day 14.
2. Human foetal livers were obtained after termination of pregnancy performed at 11–20 weeks of gestation, and with the informed consent of mothers.
3. Dissection kit.
4. Isolation medium: Dulbecco's Modified Eagle's Medium/ NUT MIX F-12 supplemented with 4 mM L-glutamine (Gibco) and 1.3 ml D-(+)-glucose solution (450 g/l, Sigma).

5. Growth medium: DMEM (Gibco) supplemented with 20% foetal bovine serum (FBS), 1% nonessential amino acid (Gibco), 2 mM L-glutamine, 0.1 mM β-mercaptoethanol (Gibco) and 1,000 U/ml leukaemia inhibitory factor (Chemicon).
6. 1× Ca^{2+}/Mg^{2+}-free phosphate-buffered saline (PBS, Gibco).
7. Collagen type IV (Sigma) from Engelbreth-Holm-Swarm mouse sarcoma. To make a stock solution, dissolve the powder in 0.25% acetic acid for several hours at 2–8°C to a concentration of 1 mg/ ml. Dilute to 10 μg/ml in PBS for coating.
8. 0.25% trypsin solution (Gibco).
9. 70 μm nylon cell strainer (BD Biosciences).
10. Cloning cylinders (glass, 8 mm×8 mm, 150 μl, Sigma).
11. Freezing medium: growth medium with 10% dimethyl sulfoxide (DMSO).

2.2. Immunocytochemistry

1. Microscope coverslips.
2. PBS.
3. TBS-T: TBS with 0.1% Triton X-100 (Sigma).
4. Fixative: 4% paraformaldehyde (PFA). 8 g of PFA was dissolved in 200 ml of PBS. The pH was adjusted to 7.0 by adding 150 μl of 10 N NaOH. The solution was stored at room temperature.
5. Blocking solution: 5% milk solution in TBS-T. Weigh out the milk powder and then add TBS-T. Mix thoroughly ensuring all of the milk has dissolved.
6. Dilution buffer for immunofluorescence staining: 10 mM HEPES pH 7.5, 0.15 M NaCl.
7. Fluorescein-labelled goat anti-rabbit IgG (H+L) (Vector Laboratories) diluted 1:200 in dilution buffer.
8. Texas red-labelled mouse anti-guinea pig IgG (H+L) (Vector Laboratories) diluted in 1:200 in dilution buffer.
9. Mounting medium with DAPI (Vector Laboratories).

2.3. Albumin Assay

1. Rat albumin ELISA quantitation kit (Bethyl Laboratories Inc., E110-125).
2. Coating buffer: 0.05 M carbonate-bicarbonate, pH 9.6.
3. Wash solution: 50 mM Tris, 0.14 M NaCl, 0.05% Tween 20, pH8.0.
4. Blocking (postcoat) solution: 50 mM Tris, 1% BSA, 0.05% Tween 20, pH8.0.
5. Sample/conjugate diluent: 50 mM Tris, 0.14 M NaCl, 1% BSA, 0.05% Tween 20, pH 8.0.

6. Enzyme substrate: TMB (3,3′,5,5′-tetramethylbenzidine, Sigma).
7. Stopping solution: 2 M H_2SO_4.

3. Methods

3.1. Tissue Culture

3.1.1. Isolation of Foetal Rat Liver Cells

1. Harvest the Fischer rat 344 embryos at embryonic day 14 (E14) by removing from the protective sac (*see* **Note 1**) and transfer them to a 100 mm Petri dish (*see* **Note 2**).
2. Add sterile PBS into a 60 mm Petri dish.
3. Under a dissection microscope, open the abdominal cavity, remove the liver tissue, which is dark pink colour under the septum transversum, and place it in sterile PBS.
4. Transfer all the liver tissues into a sterile 15 ml centrifuge tube containing 3 ml of PBS (*see* **Note 3**). Add 1 ml 0.25% trypsin solution.
5. Mix thoroughly by gently flicking the tube. Incubate at 37°C for 30 min with a gentle mixture every 10 min.
6. After digestion, add 6 ml isolation medium to inactivate trypsin and pellet the cells by centrifugation at 100×*g* for 5 min.
7. Aspirate the supernatant and resuspend the pelleted cells in 10 ml isolation medium. Mix the cells by gently pipetting up and down 5–6 times. Centrifuge at 100×*g* for another 5 min.
8. Remove the supernatant and add 10 ml isolation medium to resuspend the cells.
9. Triturate the cell solution 6–8 times gently with a flame-polished Pasteur pipette to make single cell suspension.
10. After any remaining large clumps of tissue have settled down, perform cell counting using a haemocytometer.
11. Place the cells in collagen IV pre-coated culture vessels at the density of 3,000 per cm^2 and culture the cells in growth medium.
12. Incubate the cells at 37°C in 5% CO_2.

3.1.2. Isolation of Human Hepatoblasts

1. Transfer the tissue on ice in medium (*see* **Notes 4 and 5**).
2. Put the tissue into a 60 mm Petri dish in the hood using a sterile forceps. Mince the tissue with a sterile surgical scalpel to pieces no larger than 1 mm^3.

3. Transfer the tissue into a 15 ml centrifuge tube. Add 3 ml PBS and 1 ml 0.25% trypsin solution.
4. Gently flick the tube to mix thoroughly. Incubate at 37°C for 30 min with a gentle mixture every 10 min.
5. After incubation, add 6 ml isolation medium and pellet the cells by centrifugation at 50×*g* for 5 min.
6. Remove the supernatant and resuspend the pellet with 10 ml isolation medium. Mix the cells by gently pipetting up and down 5–6 times. Centrifuge at 50×*g* for 5 min.
7. Aspirate the supernatant and add 10 ml isolation medium to resuspend the cells.
8. Triturate the cell solution 6–8 times gently with a flame-polished Pasteur pipette to make single cell suspension.
9. Filter the cell suspension using a 70 μm cell strainer.
10. Perform cell counting using a haemocytometer.
11. Place the cells in collagen IV pre-coated culture vessels at the density of 20,000 per cm^2 and culture the cells in the growth medium.
12. Incubate the cells at 37°C in 5% CO_2. Check the cells under an invert phase-contrast microscope every day (**Fig. 15.1**).

3.1.3. Cloning of Foetal Liver Progenitor Cells

1. To clone the cells with cloning ring (*see* **Note 5**), cells are first dissociated using trypsin followed by centrifugation (*see* **Section 3.1.4**).
2. Resuspend the cells with growth medium. Use a flame-polished Pasteur pipette to generate a single cell suspension.
3. Perform cell counting.
4. Seed the dissociated cells on the collagen IV pre-coated 6-well plates at 5,000, 2,000, 1,000, 500, 200 and 100 cells/well

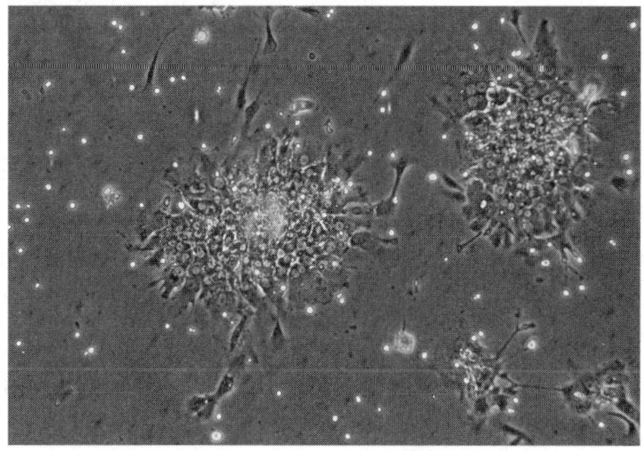

Fig. 15.1. Appearance of human foetal hepatoblasts 1 day after isolation. Phase-contrast micrograph (magnification, ×100).

(*see* **Note 6**). Culture the cells in growth medium at 37°C in 5% CO_2.

5. Inspect the cells under the inverted microscope 1–3 days after plating and a 2-day interval thereafter.
6. Mark the position of colonies that appear to have arisen from a single cell and are well separated from other cells.
7. Place a cloning cylinder with silicone grease at the bottom around a marked colony and press it down using sterile forceps (*see* **Note** 7and **8**).
8. Remove the medium inside the cloning cylinder and wash the colony once with 150 µl sterile PBS.
9. Remove PBS and add 100 µl of 0.25% trypsin solution and leave on the cells for a few seconds. Remove most of the trypsin solution and incubate the plate at 37°C. Inspect the cells periodically to monitor the detachment process.
10. When the cells are detached from the substrate, add 150 µl of growth medium to the cloning cylinder to neutralise the trypsin. Resuspend the cells by pipetting up and down gently.
11. Transfer the cell suspension to a new well of a collagen IV pre-coated 6-well plate (*see* **Note 9**).
12. Subculture the cells into a 25-cm^2 flask when the cells reach 80% confluence, and subsequently into a 75-cm^2 flask.
13. For long-term storage, the cloned cells can be frozen in liquid nitrogen (*see* **Section 3.1.5**).

3.1.4. Maintenance and Subculture

1. Maintain the cells in growth medium. Change the medium every other day.
2. Check the cells under an invert phase-contrast microscope every day.
3. Passage the cells when they reach 90% confluence (*see* **Note 10**).
4. Remove the culture medium and wash the cells with 10 ml (for a T75 flask) Ca^{2+}/Mg^{2+}-free PBS.
5. Remove PBS and add 3 ml 0.25% trypsin solution. Put the flask back to the incubator for 5 min.
6. Monitor progress of dissociation under an invert phase-contrast microscope.
7. When the cells are released from the substrate, add 7 ml growth medium to inhibit trypsin.
8. Pellet the cells by centrifugation at 100×*g* for 5 min.
9. Resuspend the cells with growth medium by pipetting up and down.
10. Aliquot the cell suspension into new culture vessels at a 1:3 ratio.

3.1.5. Storage of Foetal Liver Cells

3.1.5.1. Freezing and Storage

1. Detach the cells as for subculture with 0.25% trypsin solution.
2. Add growth medium and centrifuge at $100\times g$ for 5 min.
3. Resuspend the cells with freezing medium at a concentration of 2×10^6 viable cells/ml.
4. Transfer 1 ml of cell suspension into each cryovial.
5. Package the vials with tissue paper and place them into a polystyrene foam box for insulation. Transfer the vials to –80°C to allow to cool down slowly.
6. The following day, transfer the vials to liquid nitrogen.

3.1.5.2. Thawing

1. Retrieve the vials from liquid nitrogen.
2. Immerse immediately in 37°C water with gentle agitation to promote rapid thawing of the cells.
3. Transfer the defrosted 1 ml cells into a 15 ml centrifuge tube. Add 9 ml growth medium drop by drop with gentle shaking.
4. Centrifuge at 100 g for 5 min. Resuspend the cells in 1 ml pre-warmed growth medium.
5. Transfer the cells to a collagen IV pre-coated 25-cm^2 tissue culture flask containing 6 ml pre-warmed growth medium and incubate at 37°C in 5% CO_2.
6. Subculture when the cells are in exponential growth phase.

3.2. Characterisation

3.2.1. Fluorescence Immunocytochemistry

1. Culture the cells on collagen IV pre-coated glass coverslips (*see* **Note 11**).
2. Aspirate the culture medium and wash the cells with PBS. Fix the cells in 4% PFA for 30 min at 4°C.
3. Aspirate PFA, wash the cells with PBS once.
4. Aspirate off the PBS. Pipette 0.5 ml of TBS into each well (*see* **Note 12**).
5. Aspirate the TBS, and repeat once.
6. Pipette 0.5 ml TBS-T and leave at room temperature for 30 min (*see* **Note 13**).
7. Aspirate off TBS-T. Add 0.5 ml of milk solution to each well and incubate for 30 min at room temperature.
8. Make up the necessary amount (0.5 ml/well) of primary antibody in milk solution on ice (*see* **Note 14**).
9. Add 450 μl of the antibody solution to each well and incubate at 4°C overnight.

10. The next day, aspirate off the primary antibody.
11. Add 0.5 ml of TBS-T to each well. Aspirate off.
12. Add 0.5 ml of TBS-T to each well. Leave for 15 min.
13. Aspirate off. Wash as above *(12)* twice more.
14. Make up 0.5 ml/well of the appropriate secondary antibody solution on ice. Add 450 μl to each well and incubate at room temperature for 1 h (*see* **Note 15**).
15. Aspirate off the secondary antibody.
16. Repeat stages 11 and 12, but use TBS instead of TBS-T.
17. Mount the coverslips with mounting medium with DAPI.
18. Examine slides for specific staining using the fluorescence microscope.

3.2.2. Albumin Assay

1. Trypsinise the cells and determine the cell number using a haemocytometer.
2. Seed the cells on collagen IV pre-coated 24-well plates at a final density of 1×10^5 cells/well in 0.5 ml growth medium.
3. Incubate the cells at 37°C in 5% CO_2 for 24 h.
4. Collect the medium from each well and measure the total volume from each well to compensate the evaporation during the culture.
5. The concentration of albumin in the collected medium can be analysed using a rat albumin ELISA quantitation kit immediately after collection or store the medium at –20°C for later use.
6. Carry out the assay in a flat-bottom 96-well plate at room temperature.
7. Coat each well with 1 μl sheep anti-rat albumin affinity purified antibody (supplied with the kit) diluted in 100 μl coating buffer for 1 h.
8. Wash the wells with wash solution for three times.
9. Add 200 μl of blocking (postcoat) solution into each well and incubate the plate for 30 min.
10. After three-time wash, transfer 100 μl of standard or 200 μl of the samples to the assigned wells and incubate for 1 h.
11. To detect the signals, dilute HRP-conjugated detection antibody (1 mg/ml, supplied with the kit) in conjugate diluent in 1:40,000 and add 100 μl of the detection antibody solution onto each well. Incubate the plate for 1 h.
12. Use TMB as the enzyme substrate. Transfer 100 μl of TMB onto each well and incubate for 30 minutes.
13. To stop the reaction, add 100 μl of 2 M H_2SO_4 to each well.
14. Read the plate by a microtiter plate reader at 450 nm.

4. Notes

1. Rats should be maintained and treated according to ethical standards and specific regulations for animal care. To remove the embryos, animals are killed by cervical dislocation and should be cleaned with disinfectant. Use sterile tools for removal of embryos.
2. Always keep the embryos on ice during transportation. Put the embryos on ice block or the fridge if not performing dissection immediately.
3. Use a 5 or 10 ml pipette to transfer the liver tissue. To prevent adhesion of liver tissues to the walls of the pipette during transfer, rinse the inner surface of the pipette once with PBS before transferring the liver tissues.
4. The use of human foetal tissue for experimental use should be governed by a license from relevant regulatory authority and subject to ethical approval.
5. Keep liver tissues on ice in Eagle's minimal essential medium (EMEM) during transportation.
6. Alternatively, dilution cloning techniques can be used. Briefly, harvest the cells by trypsinisation followed by centrifugation at $100\times g$ for 5 min. Resuspend the cells gently and dilute the cells to a concentration of 10 cells/ml. Plate 100 μl of the cells into each well of a collagen pre-coated 96-well plate. Culture the cells in growth medium.
7. Use at least three different cell densities to ensure that optimal sparse cultures can be obtained. Note that sparse cultures grow less efficiently and more slowly.
8. Make sure that the colony is right in the centre of the ring. When the colony is already larger than the cloning ring, the ring may be placed over some of the cells. For the large colonies, make sure that they derive from single cells on the basis of daily observation. In some cases, large colonies composed of cells from different close colonies thus are not monoclonal.
9. Sometimes it is better to culture the cells in 24-well plates to enhance cell growth by low dilution time.
10. Do not allow the cells to form a confluent cell layer. In such a case, the cells are difficult to detach due to the stabilising influence of the adjacent cells. And confluency usually leads to morphology change of the cells, hence the phenotypic alteration, including the enhanced production of extracellular matrix, increased doubling time, etc.
11. Do not allow the cells to exceed 80% confluence since overcrowding of cells makes it difficult to distinguish cell morphology.

12. PBS buffer may be used instead of TBS buffer for all proteins except for phosphorylated proteins.
13. TBS-T buffer contains Triton, which is used to permeabilise the cell membrane in order to examine intra-cellular markers. However, to examine cell surface markers alone, TBS buffer (without Triton) may be used in place of TBS-T throughout the protocol.
14. Negative controls must be established for each secondary antibody. This is achieved by staining one coverslip each with only the secondary antibody without addition of the corresponding primary antibody.
15. Following the addition of fluorescent secondary antibody, avoid the exposure of coverslips to light to prevent degradation of fluorescence. Wrap the culture plates with aluminium foil and perform washes of the coverslips in the dark.

References

1. Fox, I. J., Chowdhury, J. R, (2004) Hepatocyte transplantation. *Am J Transpl* 4, 7–13.
2. Hughes, R. D., Mitry, R. R., Dhawan, A. (2005) Hepatocyte transplantation for metabolic liver disease: UK experience. *J Roy Soc Med* 98, 341–345.
3. Horslen, S .P., Fox, I. J. (2004) Hepatocyte transplantation. *Transplantation* 77, 1481–1486.
4. Mizuguchi, T., Mitaka, T., Katsuramaki, T., et al. (2005) *J Hepatobiliary Pancreat Surg* 12, 378–385.
5. Fausto, N. (2000) Liver regeneration. *J Hepatol* 32, 19–31.
6. Bucher, N. L., Robinson, G. S., Farmer, S. R. (1990) Effects of extracellular matrix on hepatocyte growth and gene expression: implications for hepatic regeneration and the repair of liver injury. *Semin Liver Dis* 10, 11–19.
7. Block, G. D., Locker, J., Bowen, W. C., et al. (1996) Population expansion, clonal growth, and specific differentiation patterns in primary cultures of hepatocytes induced by HGF/SF, EGF and TGF alpha in a chemically defined (HGM) medium. *J Cell Biol* 132, 1133–1149.
8. Runge, D., Michalopoulos, G. K., Strom, S. C., et al. (2000) Recent advances in human hepatocyte culture systems. *Biochem Biophys Res Commun* 274, 1–3.
9. Reid, L. M., Jefferson, D. M. (1984) Culturing hepatocytes and other differentiated cells. *Hepatology* 4, 548–559.
10. Michalopoulos, G. K., DeFrances, M. C. (1997) Liver regeneration. *Science* 276, 60–66.
11. Azuma, H., Hirose, T., Fujii, H., et al. (2003) Enrichement of hepatic progenitor cells from adult mouse liver. *Hepatology* 37, 1385–1394.
12. Williams, G. M., Weisburger, E. K., Weisburger, J. H. (1971) Isolation and long-term culture of epithelial-like cells from rat liver. *Exp Cell Res* 69, 106–112.
13. Grisham, J. W. (1983) Cell types in rat liver: their identification and isolation. *Mol Cell Biochem* 53, 23–33.
14. Neupert, G., Langbein, L., Karsten, U. (1987) Characterization of established epithelioid cell lines derived from rat liver: expression of cytokeratin filaments. *Exp Pathol* 31, 161–167.
15. Williams, G. M. (1976) Primary and long-term culture of adult rat liver epithelial cells. *Method Cell Biol* 14, 357–364.
16. Tsuchiya, A., Heike, T., Fujino, H., et al. (2005) Long-term extensive expansion of mouse hepatic stem/progenitor cells in a novel serum-free culture system. *Gastroenterology* 128, 2089–2104.
17. Tsuchiya, A., Heike, T., Baba, S., et al. (2007) Long-term culture of postnatal mouse hepatic stem/progenitor cells and their relative developmental hierarchy. *Stem Cells* 25, 895–902.
18. Yovchev, M. I., Grozdanov, P. N., Joseph, B., et al. (2007) Novel hepatic progenitor cell surface markers in the adult rat liver. *Hepatology* 45, 139–149.

19. Furukawa, K., Shimada, T., England, P., et al. (1987) Enrichment and characterization of clonogenic epithelial cells from adult rat liver and initiation of epithelial cell strains. *In Vitro Cell Dev Biol I* 23, 339–348.
20. Mahieu-Caputo, D., Allain, J. E., Branger, J., et al. (2004) Repopulation of athymic mouse liver by cryopreserved early human fetal hepatoblasts. *Hum Gene Ther* 15, 1219–1228.
21. Dan, Y. Y., Riehle, K. J., Lazaro, C., et al. (2006) Isolation of multipotent progenitor cells from human fetal liver capable of differentiating into liver and mesenchymal lineages. *Proc Natl Acad Sci USA* 103, 9912–9917.

Chapter 16

Human Intrahepatic Biliary Epithelial Cell Lineages: Studies In Vitro

Ruth Joplin and Stivelia Kachilele

Abstract

The human intrahepatic biliary epithelium is composed of a morphologically heterogeneous population of epithelial cells. During liver cirrhosis, new biliary ductular structures develop at the portal margins that express markers of immaturity such as CD56 and Bcl-2. These markers are also expressed transiently on immature biliary duct precursors during embryological development; thus their reappearance during cirrhosis suggests a recapitulation of ontogenesis during some liver conditions. Here we describe methods, based on the differential expression of membrane markers, for separating immature biliary epithelial cells from those associated with mature ducts. We also describe two- and three-dimensional culture models for the maintenance of mature and immature populations in vitro. Both populations readily establish colonies in monolayer culture but only cells from mature ducts can be maintained in medium-term culture as serially proliferating, passageable cultures; immature cells deteriorate and detach within 2–3 weeks of isolation. In three-dimensional collagen gel culture, both mature and immature populations form duct-like structures with clearly definable lumena that persist for up to 6 weeks.

Keywords: Human intrahepatic biliary epithelial cell sub-populations, purification, culture.

1. Introduction

The intrahepatic biliary epithelium lines the system of conduits that extend from the Canals of Hering (the smallest ramifications of the biliary tree) to the right and left hepatic ducts that unite to form the common hepatic duct at the porta hepatis *(1)*. The biliary epithelium is morphologically a heterogeneous tissue, showing a gradation from small and somewhat flattened cells in the smallest calibre ductules through cuboidal in interlobular ducts to tall and columnar in larger septal/segmental ducts *(1,2)*. Morphological heterogeneity of biliary epithelial cells (BEC, or cholangiocytes) is reflected in phenotypic and functional heterogeneity, particularly

Anil Dhawan, Robin D. Hughes (eds.), *Hepatocyte Transplantation, vol. 481*

DOI 10.1007/978-1-59745-201-4_16 Springerprotocols.com

in some human liver diseases *(2, 3, 4)*. Investigators with rodent models have successfully separated different cholangiocyte sub-populations based on morphological criteria *(2)*. We have utilised the differential expression of membrane markers by different subsets of biliary epithelium to purify BEC sub-populations *(4)*. We have also developed two- and three-dimensional culture models that allow us to monitor lineage progression of mature and immature BEC (iBEC) populations *(5)*.

Here we describe methods for the immunomagnetic purification of separate mature and iBEC populations from the human liver. Cells are purified by a sequence of enzymatic digestions, differential density centrifugation and immunological stages. We also describe methods for maintenance of the separated populations in two- and three-dimensional culture models.

2. Materials

2.1. Separation of Purified Mature and Immature Human Intrahepatic BEC Populations

1. Autoclave sterilised equipment: 250 ml glass beakers, 0.5 mm mesh metal sieves (Sigma), glass (fine bore) Pasteur pipettes.
2. Tissue culture plastic ware; sterile large tissue culture dishes, 10 ml and 15 ml conical base centrifuge tubes, 25 ml Universal centrifuge tubes, plastic pipettes, plastic (wide bore) Pasteur pipettes, 10 ml syringes, 0.2 um syringe filters, scalpels.
3. Stock solutions: autoclave sterilised phosphate-buffered saline (PBS – normal 1× and hypertonic 10× strengths prepared from commercially available PBS tablets).
4. Collagenase type 1A (Sigma) dissolved in normal PBS (10 mg/ml stock solution), stored as 5 ml aliquots at –20°C.
5. Normal PBS, 1% w/v bovine serum albumin (PBS, 1% BSA – pH 7.4, filter sterilised).
6. Normal PBS, 0.1% w/v bovine serum albumin (PBS, 0.1% BSA – pH 7.4, filter sterilised).
7. 0.2 M Tris, 0.1% BSA (0.2 M Tris-HCl, 0.1% BSA – pH 8.5, filter sterilised).
8. Percoll density gradient media (Pharmacia Amersham Biotech) prepared as follows: Percoll stock solution; 99 mls Percoll combined with 11 mls hypertonic (10×) PBS stored at 4°C in glass bottles; 33% Percoll solution (1.04 mg/ml); 33 mls percoll stock solution combined with 67 mls normal (1×) PBS; 77% Percoll solution (1.09 mg/ml); 77 mls of Percoll stock solution combined with 23 mls normal PBS.
9. Magnetic beads: HEA125-conjugated microbeads (Miltenyi); CD56 Dynabeads prepared by conjugation of Dynal M-450

tosylactivated Dynabeads with anti-CD56 antibody (purchased from Dako – *see* **Section 3.3**).

10. Magnetic particle concentrators: Magnetic particle concentrator (Dynal, UK); MACS separator and large cell separation columns (Mitenyi).

2.2. Monolayer Culture of Human BECs

1. Tissue culture-treated plastic 25 cm^3 flasks.
2. BEC plating medium composed of Dulbecco's Modification of Eagle's Medium (DMEM) 43.5% and Hams F12 medium 43.5% (both from Gibco) supplemented with 10% foetal bovine serum heat inactivated for 45 min at 56°C (HiFBS or human serum can be used) and the following: 0.2 M glutamine, 5 μg/ml insulin, 400 ng/ml hydrocortisone, 10 ng/ml cholera toxin, 10 ng/ml epidermal growth factor (all from Sigma) and antibiotics (10^4 IU/ml each of penicillin and streptomycin).
3. BEC growth medium is of the same composition as plating medium (point 2 above) but with only 5% HiFBS and supplemented with 10 ng/ml recombinant human hepatocyte growth factor (HGF).
4. Cell detachment solution for sub-culturing BEC; 0.25% trypsin/EDTA (Gibco), stored as 5 ml aliquots at –20°C.
5. Cryopreservation solution; dimethylsulphoxide (DMSO; Sigma and stored at room temperature). A fresh 20% DMSO solution is made at each use by appropriate dilution of DMSO in DMEM.

2.3. Three-Dimensional Culture of Human BECs in Collagen Gel

1. Type 1 collagen prepared from rats tails (*see* **Section 3.5**).
2. 0.1 % acetic acid.
3. 10× concentrated DMEM (Gibco).
4. 1 M NaOH; 0.1 M NaOH.
5. BEC plating medium (2.2 item 2 above).
6. Williams E Medium (Gibco).

3. Methods

Chronically diseased human liver (and experimentally damaged rodent liver) frequently contains numerous small biliary ductules (termed reactive ductules) that appear to arise as a general response to cirrhosis. CD56 (neural cell adhesion molecule) is a membrane molecule transiently expressed on biliary epithelium of the ductal plate during embryological development and thus represents an indicator of immature phenotype (*5,6*). In cirrhotic disorders, reactive ductules (but not mature ducts) express CD56 but lack certain other maturation markers (*5, 7*); biliary ducts and ductules in normal liver are negative for CD56 (CD56–ve). Thus reactive ductular cells in diseased human liver appear to undergo a recapitulation of ontogenesis, by expressing markers of immaturity. We have exploited this property of immature cells in diseased human liver to devise methods for purifying mature and immature populations of human BECs.

Human liver tissue is obtained through our liver transplantation programme. Approximately 30 g liver tissue (*see* **Note 1**) is obtained with informed consent: (1) from patients undergoing orthotopic liver transplantation for end-stage liver disorders (*see* **Note 2**); (2) from donor liver residual following graft size reduction for transplantation to paediatric recipients (*see* **Note 3**). Donor organs for transplantation are perfused with University of Wisconsin preservation fluid and maintained on ice; diseased liver explant samples are rapidly transferred to preservation fluid on ice. 30 g (approximately) liver slices (both donor and explanted diseased organs) are obtained and immersed in DMEM tissue culture medium and stored at 4°C until use (*see* **Note 4**). All procedures are undertaken in a sterile laminar flow cabinet unless otherwise stated. All centrifugations are at 800×*g* for 10 min with 3 min break unless otherwise stated (standard centrifugation).

3.1. Purification of Human Intrahepatic BECs

3.1.1. Preparation of Liver Tissue Extract

1. A total of 30 g of liver tissue is finely diced (dice no less than 1 mm^3) using a pair of scalpels (*see* **Note 5**) and transferred to a 250 ml sterile glass beaker containing 45 mls of normal (1×) PBS.
2. A 5 ml aliquot of collagenase type 1A is sterilised by passing through a 0.2 μm sterile syringe filter and added to the tissue dice. The dice is stirred to give a 2 mg/ml final collagenase concentration and incubated at 37°C without stirring or shaking (*see* **Note 5**) for 1–2 h (*see* **Note 6**).
3. The beaker is removed from the incubator to a laminar flow cabinet and the digest sieved through a sterile 0.5 mm mesh metal sieve. The liquor containing detached cells is collected into a clean 250 ml beaker. Tissue pieces remaining in the sieve are gently manipulated using the plunger of a 10 ml syringe to encourage further release of cells.
4. Residual undigested dice are transferred to a clean tissue culture dish and diced further with scalpels to encourage release of cells from the cut surfaces (*see* **Note 7**). The dice and sieved liquor are recombined and returned to the incubator for a further 30–60 min.
5. The contents of the beaker are strained through a clean sterile 0.5 mm mesh metal sieve and the liquor containing detached

cells collected into a 250 ml clean sterile beaker. The undigested tissue pieces are gently manipulated using the plunger of a 10 ml syringe while adding normal PBS to enhance the sieving process and wash further cells through the sieve and into a total liquor volume of 200 mls.

6. A volume of 25 ml aliquots of liquor are decanted into eight universal centrifuge tubes and centrifuged at 800×*g* for 10 min, break 3 min (standard centrifugation).
7. Supernatants are decanted and the pellets resuspended thoroughly (*see* **Note 8**) combing two pellets into a single tube (to generate four tubes) and standard centrifugation repeated.
8. Step 7 is repeated until all the pellets (liver extract pellet) are combined in one universal tube in a volume of 24 mls.

3.1.2. Semi-purification of BEC on Percoll Density Gradient Media

1. 3.5 ml of 33% Percoll is transferred into each of eight 10 ml centrifuge tubes.
2. 3.5 ml of 77% Percoll is added by inserting the pipette tip below the 33% Percoll and pipetting the 77% Percoll beneath it very slowly and gently to minimise turbulence and mixing of Percoll densities (*see* **Note 9**). This procedure results in a Percoll bi-layer with 33% Percoll floating on top of 77% Percoll (*see* **Fig. 16.1**).
3. Each of the eight Percoll gradients are gently overlayed with 3–4 ml liver extract using a plastic Pasteur pipette (*see* **Note 10**).
4. The tubes are centrifuged at 800×*g* for 30 min with no break, resulting in several discernible layers beneath the supernatant, in descending order these being: (1) a viscous floating pellicle, partially characterised as containing hepatocytes, stellate cells, undetermined non-viable cells and sub-cellular debris; (2) mononuclear cells including iBEC; (3) a range of mononuclear cells partially characterised as containing endothelium, immune cells and BEC (approximately 10%); (4) 77% Percoll, cell content undetermined (no cells detected); and (5) erythrocye pellet (*see* **Fig. 16.1**).
5. From each Percoll gradient, the supernatant and layer 1 are gently removed using a plastic pipette and discarded to waste.
6. Layers 2 and 3 (approximately 3–4 ml) are collected from each gradient and transferred to a conical base tube thus yielding eight tubes containing approximately 3–4 ml each (*see* **Note 11**).
7. The volume in each conical tube is adjusted to 10 ml by the addition of normal PBS to dilute the Percoll and the cell suspension is then standard centrifuged.
8. Supernatants are decanted and the pellets resuspended in 0.5 ml normal PBS. The pellets are combined in a single tube

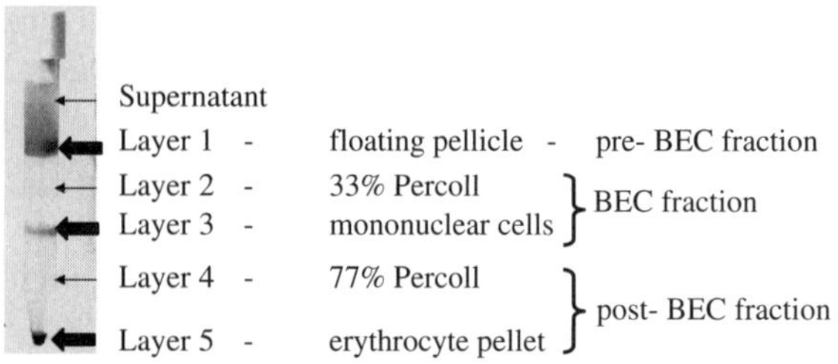

Fig. 16.1. The appearance of a Percoll gradient following centrifugation at 800×*g* for 30 min is shown. Layers 2 and 3 contain biliary epithelial cells (approximately 10%) and are harvested for further purification of immature and mBEC populations by immunomagnetic separation. The supernatant and fractions 1 and 4–6 are discarded. (*see* Color Plate 14)

and the original tubes washed thoroughly with normal PBS using a fine bore glass Pasteur pipette to detach any residual cells from the tube wall (*see* **Note 12**).

9. The volume of cell suspension is adjusted to 10 ml and standard centrifuged yielding a cell pellet containing BECs (pellet).

3.2. Immunomagnetic Separation of Mature and Immature Human Intrahepatic BEC Populations According to Differential Expression of CD56 by Immature Cells

Separate iBEC and mature BEC (mBEC) populations are purified according to the differential expression of CD56 on BEC as described by Fabris et al *(4)*; here we describe recent minor modifications to the magnetic separation stages of the process (*see* **Note 13**). Briefly, semi-purified cells are obtained by differential density centrifugation of liver homogenate on a Percoll gradient (as described in **Section 3.2** above). The total BEC fraction is purified using a pan BEC membrane molecule, epithelial glycoprotein 34, which is recognised by the monoclonal antibody HEA125 *(8)*. Using a HEA125-Miltenyi microbeads system, BEC from all subsets of bile ducts is obtained (*see* **Note 13**). Immature cells are then purified from the HEA125-positive (HEA125+ve) population by positive selection of CD56+ve (conjugated to dynabeads) cells. Thus CD56/HEA125 double-positive (iBEC) and CD56–ve/HEA125+ve (mBEC) populations are obtained.

3.2.1. Preparation of CD56-Conjugated Dynabeads

1. At least 2 days in advance of cell isolations 4×10^8 tosylactivated Dynabeads are washed 3× in normal PBS, harvesting after each wash by placing on a Dynal magnetic particle concentrator for 2 min.
2. After the final wash the beads are combined with 500 µl anti-CD56 antibody and rotated slowly for 10 min at 37°C before addition of 50 µl of 1% PBS/BSA to give a final 0.1% BSA concentration.
3. Incubation is continued for a further 24 h at 37°C with slow magnetic stirring.
4. The labelled beads are harvested on a Dynal magnetic particle concentrator for 2 min and washed 4×; washes 1 and 2 in

PBS/0.1% BSA, 5 min each at 4°C; wash thrice in Tris/0.1% BSA for 24 h at room temperature; wash four times in 0.1% PBS/BSA for 5 min at 4°C.

5. After the final wash CD56-conjugated beads are stored in PBS/0.1% BSA at 4°C (*see* **Note 14**).
6. Before use in cell separations an aliquot of the CD56-conjugated Dynabeads (10^7 beads per cell preparation) is washed in normal PBS and resuspended in 0.5 ml normal PBS.

3.2.2. Separation of Mature and Immature Human Intrahepatic BEC Populations

1. Purification of HEA125+ve cells. 40 μl HEA 125-labelled Miltenyi microbeads are added to the BEC pellet obtained from layers 2 and 3 following semi-purification by differential density centrifugation on Percoll (**Section 3.2**, step 9, above).
2. After thorough but gentle mixing, cells and beads are incubated at 4°C (or on ice) for 30 min (*see* **Note 15**).
3. During the incubation (point 2 above), two large cell separation columns are placed in the magnetic field of a MACS separator (Miltenyi) and conditioned by running 2×1 ml of normal PBS (the suspension buffer of the cells) through the column.
4. After incubation with HEA125 microbeads the volume of the cell per bead suspension is adjusted to 15 ml with normal PBS.
5. A volume of 7.5 ml of cell per bead suspension is added to each of the two columns and cell separation is performed (Miltenyi MACS). HEA125–ve cells elute into collection tubes below the columns while HEA125+ve cells are retained by the magnet.
6. Columns on the magnetic cell separator are washed 3× with normal PBS, then removed and HEA125+ve cell per bead complexes eluted into collection tubes ensuring that all cells are eluted by using the column plunger to propel the cells down the column (*see* **Note 16**).
7. The cell per bead complexes are resuspended in 30 ml normal PBS and distributed in 2×15 ml conical base centrifuge tubes and standard centrifuged for a final wash.
8. Purification of CD56+ve cells from the HEA125+ve fraction. Supernatants are decanted and the pellet containing cell per bead complexes resuspended in 0.5 ml normal PBS containing 10^7 CD56-conjugated Dynabeads (*see* **Section 3.3.1** above) and incubated at 4°C for 30 min with occasional agitation of the tube (*see* **Note 15**).
9. The cells and beads are washed in 5 ml of cold normal PBS and placed on a Dynal magnetic particle concentrator for 2 min after which the supernatant is gently decanted and retained. All residual supernatant is removed with a glass Pasteur pipette and retained with the decanted supernatant (*see* **Note 16**).

10. Normal PBS is added to the CD56+ve cells per dynabead and washing (Step 9) is repeated to a total of three repeats after which the CD56+ve per HEA125+ve cells (iBEC) are ready for plating.
11. The decanted supernatants are standard centrifuged to yield a HEA125+ve per CD56–ve pellet (mBEC).
12. The cell per bead preparations is suspended in culture medium appropriate for subsequent culture and plated (please see below).

3.3. Monolayer Culture of Human BECs

Purified mature and iBEC are cultured separately as monolayers according to our standard protocol *(5)*.

1. The entire iBEC per mBEC preparations are seeded separately into 25 cm^2 tissue culture flasks in 5 ml of plating medium and allowed to adhere for 48 h at 37°C, 5% CO_2 in air (CO_2 incubator) (*see* **Note 17**).
2. Plating medium is replaced by 5 ml of growth medium and culture continued with full medium replacement every 2–3 days until the cells become confluent (*see* **Note 18**).
3. Once confluent mBEC are harvested for subculture (*see* **Note 19** and **Fig. 16.2**) by incubating with trypsin per EDTA, the cells are washed thoroughly with normal PBS to remove all serum, and 1 ml of trypsin per EDTA is added. The cells are incubated at 37°C for 2–3 min with occasional agitation of the flask. Cell detachment is confirmed by phase contrast microscopy and 1 ml of HiFBS added to inactivate the trypsin and prevent further (damaging) digestive activity by trypsin and the detached cells are washed by standard centrifugation in 25 ml DMEM.
4. For subculture the pellet is resuspended in 10 ml plating medium for passage into 2×25 cm^2 flasks (*see* **Note 19**). Cells are left to adhere overnight and plating medium is replaced by growth medium the following day and culture continued as described above (point 2 above).
5. mBEC can be successfully frozen in and retrieved from liquid nitrogen following the second subculture (*see* **Note 20**). After detachment with trypisn and standard centrifugation (step 3 above), the cell pellet is resuspended at 2×10^6 cells per ml in cold HiFBS and an equal volume of 20% DMSO is added dropwise. The cells are rapidly distributed to freezing vials at 10^6 cells per vial (1 ml aliquots) and frozen in the vapour phase of a liquid nitrogen tank overnight before transfer to the liquid phase.
6. For retrieval of mBEC from liquid nitrogen, aliquots are rapidly thawed by placing the vials in a beaker of water at 39°C and quickly transferring the thawed vials asceptically to a laminar flow cabinet. The thawed cells are transferred to 25 ml DMEM to dilute the DMSO and standard

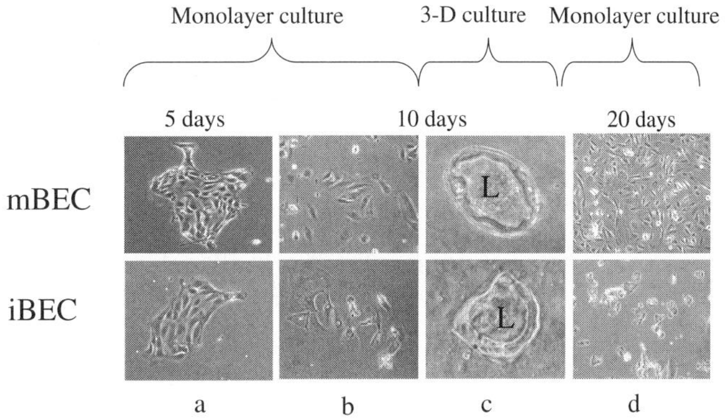

Fig. 16.2. The morphological appearance of separated immature BEC (iBEC) and mBEC fractions maintained in vitro is shown. Freshly isolated mBEC cell clusters are generally larger than iBEC clusters (around 20–100 cells vs 2–20 cells, respectively). Initially in monolayer culture iBEC and mBEC cell clusters are indistinguishable (*panel a*), but whereas mBEC develop proliferating colonies (*panel b*) that eventually become confluent (*panel d*) iBEC fail to establish and deteriorate and detach. In three-dimensional collagen gel culture, hollow spherical and ductular structures develop from both iBEC and mBEC; these structures have a clear lumen (L) surrounded by polarised epithelium (*panel c*).

centrifuged. The culture is then re-established according to Step 4 above.

3.4. Preparation of Collagen

1. Six rats' tails are immersed in sterile normal PBS for 30 min then stripped of connective tissue using pliers.
2. Excess PBS is drained and the connective tissue weighed.
3. 100 ml of 0.1% acetic acid are added per gram of wet weight of connective tissue and stirred in a sterile beaker for 2 days at 4°C after which the solution is centrifuged at 800×*g* for 30 min.
4. The pellet is discarded and the supernatant sieved through a sterile fine nylon mesh and stored at 4°C until use (up to a year).
5. Aliquots of sieved supernatant are tested for sterility before use in cell culture by placing at 37°C for 1 week and analysing for microbial growth.

3.5. Culture of BEC in Three-Dimensional Collagen Gel

1. On ice 900 μl of collagen is combined with 100 μl 10× DMEM in a universal tube and mixed thoroughly; this gives an acidic solution with a yellow indicator colour.
2. 1 M NaOH is added dropwise with thorough mixing between drops until a neutral indicator colour is approached (*see* **Note 21**). 0.1 M NaOH is then added dropwise with thorough mixing between drops until a neutral indicator colour is achieved.
3. A volume of 250 μl collagen solution is added to the central wells of a 24-well plate and allowed to gel at room temperature for 15 min. A volume of 1 ml of normal PBS is added to the outer wells of the plate to prevent dehydration of collagen in the inner wells.

4. Once gelling has occurred iBEC per mBEC preparations are seeded onto the top of the gels at approximately 10^4 per well (*see* **Note 22**) in 300 μl of plating medium.
5. The plates are incubated at 37°C in 5% CO_2 in air for 48 h after which the gels are carefully but thoroughly washed with normal PBS (*see* **Note 23**).
6. After washing the cells are overlaid with 250 μl of collagen by repeating steps 1–3 above.
7. After gelling of the second collagen layer 300 μl serum-free Williams E culture medium are added and culture is continued with medium changing every 2–3 days (*see* **Note 24**).

4. Notes

1. Although we routinely use 30 g approximately of liver tissue, it is possible to obtain adequate cell yield for further studies with as little as 5 g of tissue. Early attempts to adapt the procedure for use with liver biopsy samples were unsuccessful although we did succeed in culturing cells from biopsy fragments through an alternative approach (please see **Ref.** *(9)* for details).
2. We have successfully prepared m and iBEC from livers of patients with liver disorders including primary biliary cirrhosis (PBC), primary sclerosing cholangitis (PSC), alcoholic liver disease (ALD), polycystic liver disease, biliary atresia and alpha 1 anti-trypsin deficiency.
3. As alternative sources of normal liver we have successfully prepared mBEC from donor liver rejected for transplantation because of steatosis and uninvolved liver residual from tumour resections. We have been unable to harvest iBEC from normal liver, confirming our previous failure to identify CD56 per HEA125 double-positive cells in normal liver *(5)*.
4. Viable cells can be obtained following storage of tissue for up to 48 h. However, yield and viability of purified cells are compromised after 24 h storage and the cell purification process should commence as soon after organ harvest as is practically possible.
5. We have experimented extensively with less time-consuming, labour-saving devices such as mincers, homogenisers, stirrers in an effort to automate the dicing process. We find that such procedures release unacceptable amounts of fibrotic material and other debris that subsequently interfere with the purification process and compromise the purity of the isolated cells.

6. The duration of enzymatic digestion depends of the degree of fibrosis in the starting tissue. For tissue harvested from organ donors and which is relatively free of fibrosis we incubate for 1 h initially while samples of fibrotic tissue such as from PBC or PSC are usually incubated for a maximum of 2 h at this stage. A further incubation of 30–60 min in digestive enzyme is beneficial in releasing more cells depending on the degree of fibrotic tissue remaining after the first incubation. Incubation with collagenase beyond 3 h is damaging to cells and counterproductive in terms of achievable yield.
7. We understand that the cells harvested are those that have been digested from the surface of the dice by the combined actions of chemical and mechanical disaggregation (collagenase and dicing). Therefore increasing the surface area exposed to collagenase through effective dicing generally is effective in maximising yield of BEC, without detrimentally adding significantly to cell debris and fibrotic material in the final product.
8. A large, thick and sometimes viscous pellet is formed at the bottom of the tubes with a cloudy supernatant. Such pellets can be difficult to disaggregate and cells may adhere to the bottom of the tube and require vigorous pipetting for adequate resuspension.
9. Good Percoll gradients can be achieved if 33% Percoll is underlayed by 77% so that the 33% Percoll layer floats on top of the 77% layer; attempts to overlay 77% Percoll on top of 33% generally lead to excessive mixing of the two concentrations.
10. Liver extract should float on top of and not mix with the 33% Percoll layer. The gentler action of a wide bore plastic pipette minimises mixing of liver extract with 33% Percoll.
11. It is important that only layers 2 and 3 are harvested; although collection of some 77% Percoll (layer 4) may be unavoidable this must be minimised. Contribution of excessive volumes of 77% to the cell suspension at this stage will result in cells failing to pellet adequately at the subsequent centrifugation, leading to poor yield of cells.
12. The pellets are quite sticky and may adhere to the tube wall. Initially we experimented with coated tubes but found that with thorough washing of the standard tube wall cell loss could be minimised and coated tubes offered no advantage.
13. In the study by Fabris et al, *(4)* we used different subclasses of antibody as the basis of separation but recently amended the process in favour of utilising two different sizes of magnetic bead: HEA125-conjugated microbeads require a high-performance Miltenyi MACS particle separator at the first separation. HEA+ve cells and microbeads are not able to

bind to the Dynal magnetic particle concentrator used for harvesting the larger beads used in the second separation. Thus two cell populations are harvested based on the size of the magnetic particle to which they are bound: (1) HEA125+ve/CD56–ve cells using HEA125 microbeads; and (2) HEA125/CD56 double-positive cells using Dynabeads. At the first incubation, with HEA125-conjugated microbeads, BEC from all subsets of the biliary tree are harvested; this total population can be used for studies in which separated sub-populations are not required.

14. CD56-conjugated Dynabeads can be stored in PBS/0.1% BSA for several months at 4°C (we do not recommend freezing). After storage for more than 2 weeks we recommend two 5 min washes in PBS/0.1% BSA before final suspension in normal PBS and use in cell purification.
15. Incubation with magnetic beads must be performed at 0–4°C to prevent phagocytic activity, which can lead to the contamination of purified cells by phagocytes that ingest beads at physiological temperatures.
16. Thorough washing of the cell per bead complexes is required for with complete removal of all cells unbound to beads, thus minimising contamination of purified HEA125+ve cells with HEA125–ve cells. Such HEA125–ve cells can become trapped in bead per cell complexes and efficient washing is essential to release them and prevent reduced purity of the HEA125+ve isolate.
17. We have found that 10% HiFBS in plating medium facilitates adhesion of the cells. Some cell clusters may attach very quickly but others take longer and we routinely allow 48 h to enable all those cells able to, to adhere; no advantage has been found in allowing more than 48 h for adhesion to occur.
18. Cell clusters spread out during the first few days of culture and increase in number. After around 7–14 days of culture some cells terminally differentiate and detach. There may be a brief period of apparent cell loss. It is important to persist with culture throughout this phase as it is followed by the establishment of colonies in mBEC preparations (*see* **Fig. 16.2**), which respond to HGF in growth medium and proliferate to confluence usually within 2–3 weeks *(10)*; iBEC usually fail to establish confluent monolayers and deteriorate after 2–3 weeks (*see* **Fig. 16.2**).
19. Cells at the first subculture are still vulnerable to terminal decline if replated too sparsely. Cells should be split two ways at the first subculture followed by 1:4–1:8 at subsequent subcultures depending on growth.
20. Cells frozen in liquid nitrogen after the first passage cannot be successfully retrieved, possibly because of the large number of magnetic beads still attached to the cells at this stage.

21. It is usually possible to add 300–500 µl of 1 M NaOH before significant increase in pH is achieved but all additions should be cautious with thorough mixing in between as the solution can very quickly become too alkaline (purple indicator colour) and need to be discarded. It is advisable to use a weaker NaOH solution as neutral pH is approached as a single drop of 1 M NaOH at this stage may result in excessive alkalinity.
22. The number of cells plated is very difficult to estimate as the cells are isolated in clusters and counting on a conventional haemocytometer is untenable. It is possible to disaggregate an aliquot of freshly isolated clusters completely by extensive digestion with trypsin and to determine cell yield (found to range from 10^3–10^6 cells). Such trypsin digestions render the cells non-viable, and thus reduce the yield of cells available for subsequent studies significantly.
23. Unattached cells and beads wash easily from the surface of monolayer cultures during routine medium changes. However, after addition of the second layer of collagen, detached cells and beads are trapped within the "sandwich". Therefore, thorough washing is essential before addition of the second collagen layer, to ensure removal of as many beads and unattached cells as possible.
24. In the absence of added growth promoters both mBEC and iBEC form duct-like structures in three-dimensional collagen matrix that show circular cross-sectional profile, polarised epithelium and well-defined lumena (*see* **Fig. 16.2**). We routinely culture iBEC and mBEC in collagen gels for periods of between 2 and 6 weeks after which the cells are analysed by immunocytochemistry or PCR. We have had no success with releasing cells from the gels in a viable state for sub-culture.

Acknowledgements

The authors thank the Children's Liver Disease Foundation for support, award no. NL 1739.

References

1. Desmet, V. J. (1985) Intrahepatic bile ducts under the lens. *J Hepatol* 1, 545–59.
2. Alpini. G., Roberts, S., Kuntz. S. M. et al. (1996) Morphological, molecular and functional heterogeneity of cholangiocytes from normal rat liver. *Gastroenterol* 110, 1636–43.
3. Alpini, G., Glaser, S. S., Ueno, Y., et al. (1998) Heterogeneity of proliferative capacity of rat chiolangiocytes after bile duct ligation. *Am J Physiol* 247, 767–75.
4. Fabris, L., Strazzabosco, M., Crosby, H., et al. (1995) Characterisation and isolation of immature atypical ductular cells co-expressing NCAM and Bcl-2 from primary cholangiopathies and ductal plate malformations. *Am J Pathol* 156, 1599–1612.

5. Ishida, Y., Smith, S., Wallace, L., et al. (2001) Ductular morphogenesis and functional polarization of normal human biliary epithelial cells in three-dimensional culture. *J Hepatol* 35, 2–9.
6. Roskams, T., van den Oord, J. J., De Vos, R., et al. (1990) Neuroendocrine features of reactive ductules in cholestatic liver disease. *Am J Pathol* 137, 1019–25.
7. Crawford, J. M. (2004) Normal and abnormal development of the biliary tree, in (Alpini, G., Alvaro, D., Marzioni, M., LeSage, G., LaRusso, N., eds.) *The Pathobiology of Biliary Epithelia*, pp. 1–27. Landes Bioscience, Georgetown, TX,
8. Momberg, F., Moldenhaur, G., Hammerling, G. H. (1987) Immunohistochemical study of the surface expression of an Mr 34000 human epithelium specific glycoprotein in normal and malignant tissues. *Cancer Res.* 47,2883–91.
9. Strain, A. J., Wallace, L, L., Joplin, R., et al. (1995) Characterization of biliary epithelial cells isolated from needle biopsies of human liver in the presence of hepatocyte growth factor. *Am J Pathol* 146, 537–45.
10. Joplin, R., Hishida, T., Tsubouchi, H., et al. (1992) Human intrahepatic biliary epithelial cells proliferate in vitro in response to human hepatocyte growth factor. *J Clin Invest* 90, 1284–89.

Chapter 17

Liver Cell Labelling with MRI Contrast Agents

Michel Modo, Thomas J. Meade, and Ragai R. Mitry

Abstract

Cell transplantation is a promising approach to improve the life of patients with liver disease. At present, however, techniques to track and visualise transplanted cells in patients are fairly limited and further development of non-invasive imaging technology is needed to advance the monitoring of liver cell grafts. Magnetic resonance imaging (MRI) is a non-invasive imaging technology that already allows the visualisation of particular cell fractions in the liver by using MR contrast agents. The use of contrast agents to pre-label liver cells prior to transplantation will potentially provide a method to identify, track and study the integration of engrafted cells non-invasively by MRI. Before this technique can find its clinical application, in vitro and pre-clinical in vivo studies need to be conducted to determine the safety and specificity of this approach.

Key words: Cell Transplant, liver, MRI, contrast agent, cellular MRI, gadolinium, iron oxide.

1. Introduction

The treatment of liver disease by cell transplantation promises to save the lives of many patients. However, one of the obstacles so far has been to identify, track and visualise the integration of grafted cells non-invasively in patients over many months. Liver cell survival and its contribution to functional improvements of the liver are therefore difficult to assess.

Non-invasive imaging is needed to probe the liver repeatedly. Imaging techniques, such as magnetic resonance imaging (MRI), have developed sophisticated probes that allow the distinct visualisation of different cell fractions in the liver *(1,2)*. Contrast agents, such as mangafodipir trisodium (Mn-DPDP, Teslascan), will integrate into all hepatocytes, whereas Gd-based agents (e.g. Gadoxetate) will only visualise mature hepatocytes. As in the case of hepatomas, this difference between probes can therefore be

Anil Dhawan, Robin D. Hughes (eds.), *Hepatocyte Transplantation, vol. 481*

DOI 10.1007/978-1-59745-201-4_17 Springerprotocols.com

exploited to determine if immature highly proliferative hepatocytes are present. In contrast, ferumoxides, such as Endorem, are taken up and processed by Kupffer cells and can consequently be used to visualise the liver's resident macrophages. The difference in cell uptake of MR contrast agents can hence also be used, for instance, to study acute liver rejection *(3)*.

Although these agents allow the specific imaging of different cell types within the liver and help to determine if a particular cell type has been replaced by grafting, these probes do not allow the distinction between grafted and non-grafted cells. For this, MRI contrast agents need to be incorporated in vitro into the cells prior to their transplantation. Several types of contrast agents based on gadolinium, manganese or iron oxide have been described for cellular tracking *(2)*. Currently, the most commonly used contrast media are ferumoxides. This is largely due to their superparamagnetic relaxation properties (i.e. the relaxivity effect is about 50 times larger than the contrast particle) that allow the detection of even small numbers of cells *(4)*. The use of micron-sized particles of iron oxide (MPIOs) even allows the visualisation of single cells by MRI *(5)*.

The development of bimodal agents, i.e. probes that can be detected by more than one imaging modality, provides a further development that is an efficient system to study the effects of contrast agents in vitro, and also allows the corroboration of in vivo imaging *(6–8)*. Although these agents provide additional benefits over currently available clinical probes, at present these agents are mainly used for experimental studies and will be required to undergo a stringent assessment prior to clinical approval. Clinically approved agents for liver imaging, such as Endorem or Teslascan, might be more readily implemented into clinical protocols as they are approved agents for liver imaging and in most cases will provide sufficient flexibility to identify grafted cells. The clinical translation of cellular MRI has recently been described with a clinical trial assessing Endorem-labelled dendritic cell placement in patients with lymphomas *(9)*, therefore providing a precedent that these agents can be safe for cellular imaging of implanted cells.

We here present the methodological framework in which the effects of MRI contrast agent incorporation in liver cells can be assessed prior to progressing to in vivo experiments.

2. Materials

1. Williams' medium E (cat no. W1878, Sigma, UK).
2. Foetal calf serum (cat no. F4135; 10%, v/v, Sigma, UK).
3. Penicillin/streptomycin (cat no. P0781; 10,000U/10 mg per ml, Sigma, UK).

4. L-Glutamine (cat no. G7513; 5 ml of 200 mM, Sigma, UK).
5. Insulin (cat no. I1884; final concentration of 0.1 μM, Sigma, UK).
6. Dexamethasone (cat no. D8893; final concentration of 0.1 μM, Sigma, UK).
7. Lipofectamine2000 (cat no. 11668, Invitrogen, UK).
8. Anti-dextran antibody (cat no. 10730, Stem Cell Technology, USA).
9. Trypan blue (cat no. 15250061, Invitrogen, UK)
10. Fluorescein diacetate (FDA) (cat no. D2650, Sigma, UK).
11. Ki67 (cat no. NCL-Ki67p, Novocastra, UK).
12. CyQuant (cat no. C7026, Invitrogen, UK).
13. 3-(4,5-Dimethylthiazol-2yl)-2,5-diphenyltetrazolium bromide, referred to as MTT (cat no. M2003, Sigma, UK).
14. [^{14}C]-Leucine (cat no. CFB183, 1 ml of 1.85 MBq; Amersham International, Buckinghamshire, UK).
15. Packard FilterMate (Packard Instruments, Berkshire, UK).
16. Packard Matrix 9600 ß-counter (Packard Instruments, Berkshire, UK).
17. (5-and 6)-Chloromethyl-2′,7′-dichlorodihydrofluorescin diacetate, acetyl ester (CM-H2DCFDA) solution (cat no. c6827, Invitrogen, UK).
18. Endorem/Feridex (Guerbet, France/Berlex, USA).
19. Gadophrin-2 (Schering, Germany).
20. Multihance (Bracco, Italy).
21. Primovist (Schering, Germany).
22. Resovist (Schering, Germany).
23. Teslascan (Amersham, USA).

3. Methods

3.1. Cell Culturing and Labelling

3.1.1. Cell Culturing

Human hepatocytes are seeded onto collagen type I-coated culture vessels, such as 96-well plates used for the functional assays, and glass coverslips for confocal microscopy. After seeding, the cultures should be incubated overnight in a humidified incubator (37°C, 5% CO_2) prior to commencing the labelling procedure.

1. Human hepatocytes are seeded at 50,000 cells in 200 μl Williams' medium E +supplements per well of collagen-coated 96-well plates.
2. The plates should be incubated overnight in a humidified incubator (37°C, 5% CO_2).
3. Cells are then labelled with the MRI contrast agent (*see* **Section 3.1.2**).

3.1.2. Labelling

The choice of contrast agent will depend on how many cells and which type of cell one intends to visualise after transplantation. Incorporation of clinically approved contrast agents that can be used for tracking grafted cells are Fe-based agents, Mn-based agents or Gd-based agents *(10)*.

Ideally, cellular MRI does not interfere with the general assessment of liver pathology. The use of alternative nuclei for MR imaging, such as ^{19}F, might therefore be exploited *(11)*. However, these approaches often suffer from low signal-to-noise ratios. If only a small number of cells are transplanted, alternative nuclei might not produce sufficient signal to allow a reliable detection.

Pinocytosis. Liver cells easily take up particular types of contrast agents. Labelling a particular cell population can therefore be facilitated by choosing the appropriate agent designed to be taken up by this type of cell. Some contrast agents with a small molecular weight, such as Gd-based agents, might also get taken up in vitro into the cells through fluid phase pinocytosis. For this:

1. Culture cells according to standard protocol.
2. Following overnight incubation of cultures, the contrast agent is added at appropriate concentrations to fresh media and gently shake the culture to ensure good mixing of the contrast agent with the media. The concentration of contrast agent will depend on the contrast agent and type of cell. Clinically approved agents generally come in a prepared solution and it is recommended to start with three sets of concentrations 1:1, 1:10 and 1:100. A further refinement of this dilution assay is needed to determine the best molar concentration for cell labelling. Knowing the molar concentration will be important to assess the relaxivity characteristics of the contrast agent. Typically, iron oxide-based agents will be in the range of μM, whereas Gd-based agents will be in the range of mM.
3. Duration of incubation. Certain cells, such as Kupffer cells, rapidly incorporate contrast agents and incubation times of <2 h can be sufficient for cell labelling. However, duration of incubation also depends on the concentration of contrast agent in the media. The advantage of bimodal agents is that during this process, it is possible to assess cell uptake of the contrast agent under an inverted fluorescent microscope.

4. After sufficient contrast agent has been incorporated into the cells, wash the cells 3× with phosphate-buffered saline (PBS) before adding media for further experimentation.

Transfection Agents. Although liver cells will easily take up various contrast agents, in some cases it might be desirable, for instance, to label hepatocytes with iron oxide particles that typically are not incorporated into these types of cells. The use of transfection agents can enable this process and ensure sufficient cellular uptake of particles to allow a reliable detection. For this:

1. Prepare transfection solution with 5 μl of Lipofectamine2000 in 25 μl culture media for each well on a 24-well plate.
2. Mix the contrast agent with transfection solution for 10 min on a shaker at room temperature. The contrast agent concentration will determine how much agent needs to be mixed with the transfection agent. A typical guidance is about 100 μg of ferumoxides to 5 μl of transfection agent.
3. Incubate for 2–3 h in a 1:1 mixture of serum-free media and transfection agent-coated contrast agents. However, specific incubation times will depend on the contrast agent, the transfection agent and the type of cells.
4. Remove supernatant and wash cells three times with PBS before adding culture media for further experimentation.

3.2. Visualisation of Contrast Agent Inside Cells by Microscopy

3.2.1. Detecting Iron Particles

Iron particles can be detected histologically by Perl's stain *(12)*. For this:

1. Prepare Perl's solution with 1.0 g of potassium hexacyanoferrate (ferrocyanide), 25 ml of distilled water, and 25 ml of 13% hydrochloric acid. This solution should be freshly prepared.
2. Wash cells or tissue with distilled water and add Perl's solution for 20–30 min
4. Wash cells or tissue with distilled water and add neutral red to the tissue sections for 1–2 min. For cells, this step can be omitted as neutral red counterstains tissues in shades of red.
5. Rinse with tap water and dehydrate with graded alcohols (70, 80 and 100%), clear and mount.
6. Ferric iron will appear blue.

However, Perl's stain detects all iron and therefore will also pick up other cells that contain iron, such as macrophages or blood cells. Ideally, this method is therefore only used in vitro or in tissues that do not have cells that naturally contain large quantities of iron. As liver tissue is generally used as a positive control for Perl's stain due to its naturally high content of iron, it is not recommended to use this method to detect iron-based contrast agents inside liver tissue.

3.2.2. Detecting Contrast Agents Based on Dextran

Many MRI contrast agents use dextran as a chelating agent and an immunocytochemical approach can therefore be used to specifically detect dextran-based contrast agent:

1. Label cells with a contrast agent.
2. Permeabilise cells or tissue with a 0.1% Triton X solution for 5 min.
3. Rinse cells per tissue with PBS.
4. Add FITC-conjugated anti-dextran antibody at 1:1000 dilution in PBS to the cells or tissue.
5. Rinse with PBS.
6. Counterstain all cell nuclei with DAPI or Hoechst.
7. The contrast agent will appear in green, whereas cell nuclei in blue.

The green fluorescent signal should be clearly localised to particles within the cells. If there is a diffuse staining or an absence of staining, this could indicate that the dextran chelate is being downgraded within the cells or that there are no particles present within this cells, respectively.

3.2.3. The Use of Bimodal Agents

The use of bimodal contrast agents facilitates the visualisation of cellular uptake. Due to the fluorescent moieties in these agents, it is possible to directly visualise the contrast agent as it is taken up into the cells under an inverted fluorescent microscope. Moreover, it is also possible to easily determine the cellular compartments within which the agent is trapped within the cells *(13)*. A bimodal agent currently undergoing clinical development consists of Gadophrin-2, which can be used to label liver cells for cellular imaging after transplantation *(8)*. Bimodality might also refer to other combinations of imaging modalities and care should therefore be taken that the appropriate imaging modalities can be used to visualise the agent of interest. Examples of bimodal contrast agents labelling liver cells are presented in **Fig. 17.1**.

3.3. Assaying the Effects of Contrast Agents on Cell Function

3.3.1. Cell Viability – FDA

It is essential to measure the viability of the cells after cell labelling. Contrast agents contain metal particles that are rendered non-toxic through the use of chelating agents, such as dextran or

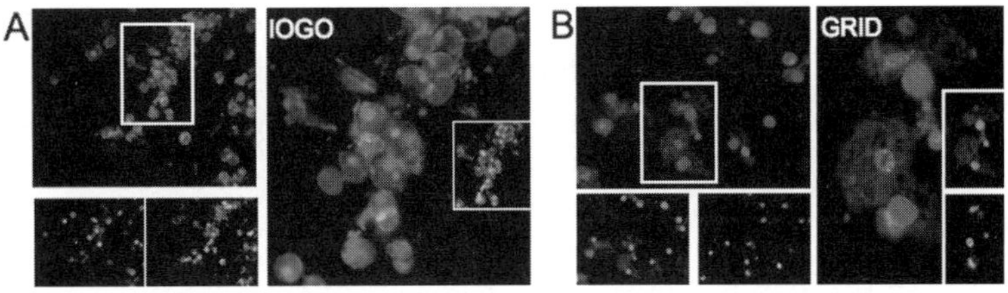

Fig. 17.1. Visualisation of the MRI contrast agent. (**A**) Adult human hepatocytes being labelled with the bimodal Iron Oxide Green Oregon (IOGO) contrast agent (in green). Note that some cells (cell nuclei in blue) are not labelled. It is noteworthy that the contrast agent seems strongly associated with the cell nuclei and does not fill the cytoplasm. It is likely that mainly phagocytic Kupffer cells incorporated this agent, whereas unlabelled cells represent a small fraction of undifferentiated hepatocytes. (**B**) In contrast, the Gadolinium Rhodamine Dextran (GRID) bimodal agent (*in red*) clearly labels the cytoplasm of cells that have the appearance of immature hepatocytes and is incorporated into all types of cells. (*see* Color Plate 15)

albumin. Upon degradation of the protective coating, these metal particles can affect cell viability. Moreover, overloading of the cells with the contrast agent can also lead to deterioration in cell viability. However, it is not sufficient to just measure cell viability straight after labelling, but ideally more protracted time points relevant to transplantation paradigms should also be investigated. A variety of cell viability assays are commercially available, such as trypan blue marking all dead cells under brightfield or FDA to label all viable cells under fluorescence microscopy.

This test requires the use of the FDA stock solution (5 mg/ml in DMSO) *(14)* and cells must adhere to glass coverslips.

1. Following labelling with an MRI contrast agent, remove the culture medium and gently rinse twice with PBS.
2. Replace PBS with 200–300 μl medium containing FDA (2 μg/ml; final concentration) and incubate the cultures for 6 min at room temperature.
3. Remove medium and gently rinse twice with PBS.
4. Counterstain cell nuclei with DAPI or Hoechst.
5. Use a fluorescence microscopy to check for green fluorescence in the cytoplasm of the viable cells only.
6. A quick semi-quantitative estimate of viable cells (green) could be carried out in e.g. five random fields. DAPI-stained nuclei could be counted as this will help in estimating the approximate number of total cells in the counted field.

3.3.2. Cell Proliferation/ Mitochondrial Activity

Incorporation of contrast agents into cells will lead to their compartmentalisation within the cells. As cells divide, the amount of contrast agent between cells will also decrease with time. However, contrast agents can also affect the cells' ability to proliferate by interfering with basic cell functions involved in mitosis. Either the counting of proliferating cells based on the number of cells in a culture dish or the use of antibodies, such as Ki67, labelling all dividing cells can be used for this. Moreover, various commercially available assays, such as CyQuant, are available that assess various aspects of proliferation and often are taken as a measure of proliferation. For instance, the MTT assay assesses mitochondrial activity that is related to mitosis.

A commonly used assay to determine the overall cell metabolic activity based on mitochondrial dehydrogenases activity consists of the MTT assay *(15)*:

1. Label cells with an MRI contrast agent.
2. Prepare the MTT assay solution with 5 mg/ml in PBS (pH adjusted to 7.2 and filtered through a 0.2 m filter). This solution can be stored in the dark at 4°C for up to 2 weeks. (*See* **Note 1.**)
3. Dilute MTT solution in culture medium (1:10) and incubate with cells for 4 h.
4. Remove media with MTT and place 20 µl of 0.25% trypsin per well in a 96-well plate and place on a shaker for 5 min at high speed.
5. Add 100 µl of isopropanol with 0.04 N HCl and place on a shaker for 15 min. This will dissolve the formazan.
6. Measure absorbance at 595–655 nm to quantify mitochondrial activity.

It is important to include appropriate control conditions (such as no MTT and no contrast agent) and express fluorescence absorbance in relation to these controls. Results based on this assay can be found in **Fig. 17.2**.

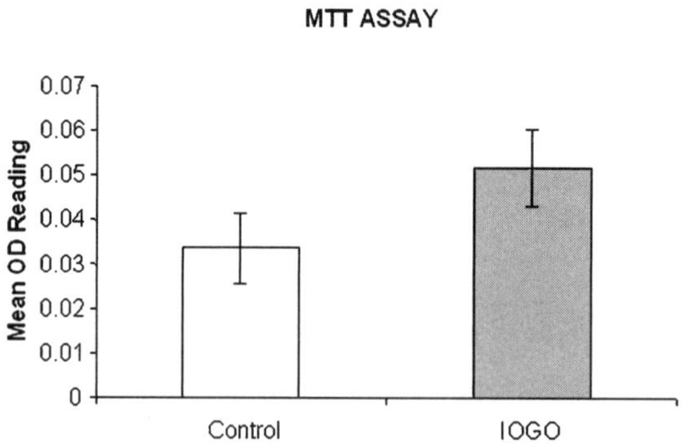

Fig. 17.2. Effects of labelling adult human hepatocytes with the Iron Oxide Green Orgeon (IOGO) MRI contrast agent on the overall cell metabolic activity compared to the control. Although the MTT is often used to measure proliferation based on mitochondrial activity involved in protein synthesis, under certain circumstances an increased activity will be a reflection of increased activity in the cells rather than proliferation. IOGO, in this case, did not increase cell proliferation, but the cells processing of the agent resulted in an increased activity.

3.3.3. Protein Synthesis – [^{14}C]-Leucine Incorporation Assay

This assay is used as an indirect functional assay that reflects the overall synthetic activity of the cells *(14)*. [^{14}C]-leucine is a radioactive amino acid that gets incorporated in proteins. Following incubation of the cell cultures, the cells are harvested onto a glass membrane and the membrane is dried followed by the radioactivity being counted using a Packard Matrix 9600 ß-counter (Packard Instruments). The counts are presented as counts per minute (c.p.m.). If the MRI contrast agent used has cytotoxic effects, it is expected to result in lower counts.

This assay requires the use of a 96-well cell culture plate. At the time of replacing the culture medium with medium containing the contrast agent:

1. Add [^{14}C]-leucine solution to the medium to give a final dilution of 0.2 μCi/well.
2. Label the plate with appropriate radioactivity warning signs.
3. Incubate the plate for the required period of labelling.
4. Post incubation, the cells are harvested and their membranes are analysed (see step 5) or the plate could be sealed with parafilm to be stored at –20°C for later analysis.
5. At the time of harvesting cells, the plate temperature should approximate room temperature, i.e. frozen plates must be completely defrosted.
6. Cells are harvested onto a glass membrane using the cell harvester.
7. The membrane must be dried in an oven (50–60°C) for 2–3 h. Ensure that the membrane is completely dry, otherwise contamination of the counter will occur.
8. Count the radioactivity of the membrane using the ß-counter for 6 min and calculate c.p.m.

High counts means a high level of protein synthesis and this will indirectly reflect the level of cellular synthetic activity (*See* **Note 2**).

3.3.4. Reactive Oxygen Species

Reactive oxygen species (ROS) should be measured in response to cell labelling to determine if the procedure produces any stressors to the cells. ROS should be measured straight after labelling and at least 24 h post-labelling to determine if these cells are undergoing a continued stress or if it is only a transient phenomenon associated with the labelling procedure rather than the presence of the contrast agent. It is possible that some of the contrast agents are degraded inside the cells, which can lead to the production of ROS and result in cell death.

Labelling of cells with an MRI contrast agent can lead to the cell undergoing reactive stress. To determine to what degree this leads to the production of ROS that can damage the cells and lead to cell death, an ROS assay can evaluate if the cell labelling might exert deleterious effects on the cells. For this:

1. Label cells with an MRI contrast agent.
2. Prepare the CM-H2DCFDA solution by diluting 50 μg in 8.6 ml of PBS.
3. Add 1 ml of this solution to each well of a 24-well plate and leave on cultures for 60 min.
4. Wash 3× with PBS prior to fixing cells with 4% paraformaldehyde.
5. Quantify green fluorescence in an FITC channel. If ROS are present, these will emit a green fluorescent light.

If no ROS are present, no green fluorescence will be emitted. It is important to include a control condition to gain an estimate of the natural background. Values can then be expressed in relation to this control condition to reflect an increase in ROS. (*See* **Note 3**).

3.4. MR Relaxometry

Magnetic resonance images (MRI) are very dependent on the sequences used to acquire the images. Sequences can be designed to highlight fluids (such as on T2-weighted images) or to be fairly insensitive to fluids (such as on T1-weighted images). The effect of contrast agents on the signal in images will depend both on the type of sequences that are being used to scan a sample and on the strength of the magnet. As field strength increases, the signal-to-noise ratio and spatial resolution increase. However, contrast agents do not necessarily follow the same principle. It is therefore important to bear these factors in mind if clinical translation is envisaged. Most pre-clinical and cell labelling experiments are conducted on high-field-strength magnets (>4.7 T), whereas most clinical studies are conducted at either 1.5T and 3T, possibly resulting in too little signal to detect transplanted cells.

To determine if sufficient contrast agent has been incorporated into the cell to effect a signal change on MRI, relaxometry needs to be conducted to quantify the relaxation signal on an MR image. For this:

1. Label cells with a contrast agent.
2. Cell are placed in a vial (e.g. Eppendorf tube) with media or PBS.
3. Comparisons should include cells with no contrast agent, media/PBS and distilled water to determine the specific change that the contrast agent induces inside the cells.
4. Insert Eppendorf into the coil of the scanner using either custom-made holders for the Eppendorfs or embed Eppendorfs into agarose gel for scanning. It is also possible to add cells directly into agarose gels and to insert these into the coils for scanning. In some cases, agarose gels are preferred as the signal of these often resembles that of tissue. Ideally, several comparisons can be run in one scanning session (a standard control should be included for all scans).

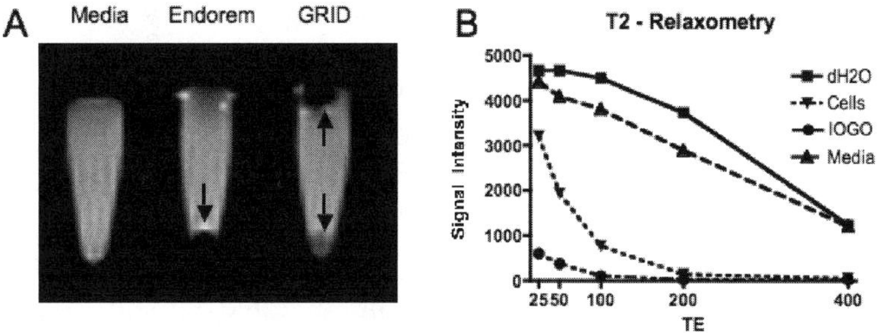

Fig. 17.3. MR relaxometry. (**A**) shows a T2-weighted MR image of Eppendorfs with media, Endorem-labelled or GRID-labelled cells (downward arrows). Air bubbles (upward arrow) on T2-weighted scans can easily be confounded with contrast agent-induced signal loss and care must be taken in the interpretation of these hypointensities. (**B**) Based on these images, it is possible to measure the signal change if the echo time is varied. It is these values that are used to calculate the relaxivity of the contrast agent. It is noteworthy here that cells clearly produce less signal than media or water. Incorporation of the bimodal agent IOGO in these cells further reduced their signal indicating the efficiency of this agent compared to only cells.

5. Scanning parameters will depend on the scanner hardware. T1 and T2 relaxivities should be measured for all conditions. It is recommended that scanning sequences for relaxometry should be set up by an experienced MR physicist.
6. On these images measure the signal in the area of the cells and the media for each echo time (for T2 relaxation) or relaxation time (for T1 relaxation). This will allow to determine the contribution of the contrast agent to the relaxivity of labelled cells and if there is leakage of the contrast agent into the media.
7. The results on the change in the signal can be plotted and analysed using a multiple regression. It is advisable to log transform the data to conduct this analysis. For measuring the relaxivity of a particular contrast agent, a 1 M solution can be used to express the molar relaxivity of the compound. However, this can also be calculated based on the molar concentration of the solution.

Calculating the relaxivity of contrast agents within cells will be essential to calculate how many cells can be detected and will provide the basis for deciding which contrast agent is more effective for identifying transplanted cells in vivo by MRI. Results based on this method are presented in **Fig. 17.3**.

3.5. Quantifying Cellular Uptake of MRI Contrast Agents

Inductively coupled plasma–mass spectrometry (ICP-MS) can be used to quantify the uptake of various contrast agents into cells. For this:

1. Label cells with a contrast agent of interest.
2. Use sufficient quantities to yield >60,000 cells in 250 μl of media. However, it is essential to count the number of cells

per condition using a hematocytometer. An exact number of cells is needed to calculate how much contrast agent was taken up per cell.

3. Ideally, include control conditions with only cells, media plus contrast agent and only media to define the background or noise in the measurements.
4. Digest the sample overnight with an equal volume of nitric acid.
5. From this digested sample, add 0.25 ml to a mixture of 0.05 ml indium (serving as internal control), 0.3 ml of nitric acid and 9.4 ml of distilled water.
6. A sample of this mixture is then sprayed into the ICP-MS (this needs to be done by an experienced researcher).
7. From these results, calculate the amount of contrast agent per cell by dividing the total amount of particles (expressed in moles or mg) in the total sample by the number of cells to yield a concentration of mol/cell or mg/cell.

Knowing the amount of mole per cell will provide the basis to calculate after how many cell divisions it will no longer be possible to detect cells by MRI, but it will also help to determine if the amount of contrast agent per cell needs to be increased to ensure a more reliable detection. This measure will also be essential to determine how effective a particular agent is to change relaxivity. If a large quantity of intracellular contrast agent is needed to effect relaxivity, it is preferable to choose a contrast agent that is more effective and would require less cellular uptake.

4. Notes

1. If precipitation occurs in the MTT solution a few days after preparation, this could be removed by filtration through a 0.2 μm filter.
2. DNA synthesis assay, which uses [^{3}H]-thymidine incorporation, would be a useful assay to determine the effects of MR contrast agent labelling on cell proliferation. This cannot be used in the case of adult human hepatocytes as they do not divide in vitro.
3. The above assays could be carried out with any type of mammalian cells or cell-lines. Cell type-specific assays could be carried out if needed, e.g. for hepatocytes, albumin (liver-specific protein) level in the cell culture supernatant or urea synthesis (liver-specific detoxification product of NH_4^+ metabolism).

Acknowledgements

The authors thank Profs Steve Williams and Jack Price for their continued support in the development of cellular MRI. MM is currently supported by a RCUK fellowship and the Wolfson Foundation.

References

1. Balci, N. C., Erturk, S. M. (2007) Cellular MR imaging of the liver using contrast agents, in (Modo, M., Bulte, J. W., ed.), *Molecular and Cellular MR Imaging*, pp. 247–258. CRC Press, Boca Raton, FL.
2. Modo, M., Hoehn, M., Bulte, J. W. (2005) Cellular MR imaging. *Mol Imaging* 4,143–164.
3. Muhler, A., Freise, C. E., Kuwatsuru, R., et al. (1993) Acute liver rejection: evaluation with cell-directed MR contrast agents in a rat transplantation model. *Radiology* 186, 139–146.
4. Bulte, J. W., Kraitchman, D. L. (2004) Monitoring cell therapy using iron oxide MR contrast agents. *Curr Pharm Biotechnol* 5, 567–584.
5. Shapiro, E. M, Sharer, K., Skrtic, S., et al. (2006) In vivo detection of single cells by MRI. *Magn Reson Med* 55, 242–249.
6. Modo, M., Cash, D., Mellodew, K., et al. (2002) Tracking transplanted stem cell migration using bifunctional, contrast agent-enhanced, magnetic resonance imaging. *Neuroimage* 17, 803–811.
7. Mulder, W. J., Koole, R., Brandwijk, R. J., et al. (2006) Quantum dots with a paramagnetic coating as a bimodal molecular imaging probe. *Nano Lett* 6, 1–6.
8. Daldrup-Link, H. E., Rudelius, M., Metz, S., et al. (2004) Cell tracking with gadophrin-2: a bifunctional contrast agent for MR imaging, optical imaging, and fluorescence microscopy. *Eur J Nucl Med Mol Imaging* 31, 1312–1321.
9. de Vries, I. J., Lesterhuis, W. J., Barentsz, J. O. , et al. (2005) Magnetic resonance tracking of dendritic cells in melanoma patients for monitoring of cellular therapy. *Nat Biotechnol* 23, 1407–1413.
10. Karabulut, N., Elmas, N. (2006) Contrast agents used in MR imaging of the liver. *Diagn Interv Radiol* 12, 22–30.
11. Ahrens, E. T., Flores, R., Xu, H. et al. (2005) In vivo imaging platform for tracking immunotherapeutic cells. *Nat Biotechnol* 23, 983–987.
12. Perls, M. (1867) Nachweis von Eisenoxyd in gewissen Pigmenten. *Virchows Archive der Pathologie, Anatomie und Physiologie* 39, 42–48.
13. Brekke, C., Morgan, S. C., Lowe, A. S., et al. (2007) The in vitro effects of a bimodal contrast agent on cellular functions and relaxometry. *NMR Biomed* 20(2), 77–89.
14. Friend, J. R., Wu, F. J., Hansen, L. K., et al. (1999) Formation and characterisation of hepatocyte spheroids, in (Morgan, J. R., Yarmush, M. L., ed.), *Methods in Molecular Medicine: Tissue Engineering Methods and Protocols,* pp. 248–249. Totowa, NJ, Humana Press Inc.
15. Mitry, R. R., Hughes, R. D., Bansal, S., et al. (2005) Effects of serum from patients with acute liver failure due to paracetamol overdose on human hepatocytes in vitro. *Transplant Proc* 37, 2391–2394.

Chapter 18

Microbiological Monitoring of Hepatocyte Isolation in the GMP Laboratory

Sharon C. Lehec

Abstract

For clinical hepatocyte transplantation, cells need to be prepared in a sterile GMP environment. Strict regulations are in place that set the standard for this environment that cells are prepared in. These regulations control all aspects of the environment. In the United Kingdom, the laboratory must have a licence from the Human Tissue Authority to prepare cell for clinical administration. The physical parameters such as air quality, pressure, temperature and microbiology counts have to be monitored regularly usually through direct measurement. Described here are the methods for microbial monitoring of the laboratory environment and the isolated cell preparations.

Key words: microbial contamination, blood culture, environment, sterility

1. Introduction

Microbial monitoring of the laboratory should be carried out weekly. This is to ensure that any potential microbial contamination is kept within prescribed limits and that the appropriate action is taken if these limits are approached or exceeded. The room air systems must be in operation and laminar flow cabinets should be on while monitoring is taking place. Microbiological monitoring of cell preparation must be performed during every cell isolation procedure *(1)*.

When setting up environmental monitoring of a laboratory, the number of sampling points needs to be decided to ensure adequate coverage. This will depend on the size of the room. A record sheet should be made to record results. It is also useful to make a diagram of the facility marking the position of the sampling points.

Anil Dhawan, Robin D. Hughes (eds.), *Hepatocyte Transplantation, vol. 481*

DOI 10.1007/978-1-59745-201-4_18 Springerprotocols.com

2. Materials

1. Tryptone Soya Agar (TSA) contact plates (Cherwell, Bicester, UK).
2. TSA settle plates (Cherwell).
3. Air sampler (F.W. Parrett Limited, London, UK).
4. BacT/ALERT bottle (BioMérieux UK Limited, Basingstoke, UK).

3. Methods

3.1. Microbial Monitoring of Laboratory

Microbiological monitoring is carried out using irradiated TSA settle plates to detect microorganisms in the air and TSA contact plates for surface contamination.

Before beginning monitoring, check the plates, do not use (a) cracked plates, (b) plates that accidentally fall open, (c) plates where the agar has been touched by fingers or the plate lid, (d) plates showing signs of microbial growth and (e) plates in which the agar has dried.

3.1.1. Settle Plate Count for Airborne Microorganisms

1. Settle plates are petri dishes containing a medium, which is usually agar-based and which will encourage and support the growth of bacteria and fungi, which land on them.
2. The purpose of the settle plate count is to monitor the cleanliness of an environment.
3. Settle plates must be inverted when being stored and incubated.
4. Collect pack of settle plates.
5. Label the bottom of the plate with the following information:
    a. The location code.
    b. The date.
6. Place the settle plates in the appropriate position, as indicated on the record sheet and diagram.
7. Expose the agar surface placing the lid face down next to the plate.
8. Plates should be exposed for a minimum of 1 h up to a maximum of 4 h.
9. Replace the lids and collect the settle plates.
10. Seal the lids with at least two pieces of fresh adhesive tape.
11. Plates should be placed in a bag and sealed.
12. Leave at room temperature for 3 days (to encourage any fungal colonies to grow).

13. Incubate at 32°C for 4 days.
14. After incubation, results should be read with number of colonies counted and recorded on the microbiology monitoring form. (*See* **Notes 1–3** for interpretation of results.)
15. Once plates have been read, they should be disposed of by autoclaving.

3.1.2. Contact Plate Count for Surface Microorganisms

1. Contact plates are agar plates that can be used to take surface samples. A contact plate is a plastic dish filled with agar to give a convex surface with an area of 25 cm^2 and can therefore be pressed against a test surface. The count of colonies after incubation can be directly related to the contamination as cfu per unit area.
2. The purpose of contact plates are:
   (i) To monitor the cleanliness of surfaces, e.g., benches, floors, hatches, etc.
   (ii) To show the effectiveness of cleaning schedules.
3. It will give a total aerobic count. The surface under test may be sampled before cleaning.
4. Collect pack of contact plates.
5. Label the bottom of the plate with the following information:
   (a) the location code
   (b) the date.
6. Samples should be taken in the appropriate position, as indicated on the record sheet and diagram.
   Sampling is carried out as follows:
   (i) Remove the lid taking care not to touch the agar surface.
   (ii) Press the agar into contact with the test surface
   (iii) Apply a firm and even pressure on the test surface for a few seconds taking care not to smear the agar over the test area.
   (iv) Replace the lid and seal with at least two pieces of fresh tape.
7. Clean the area that has been sampled with an alcohol wipe and sterile 70% IMS.
8. Collect the contact plates.
9. Plates should be placed in a bag and sealed.
10. Leave plates at room temperature for 3 days.
11. Incubate at 32°C for 4 days.

3.1.3. Air Sampling

Once a month, the air quality of the unit is tested at various locations in the Cell Isolation Unit, to ensure that aseptic processing can be performed. This is done by taking 1 m^3 sample of air using an air sampler that draws a measured sample of air onto an agar plate.

1. Ensure that the air sampler is clean before use with alcohol wipes and sterile 70% IMS, paying particular attention to the head where the sample is taken. Allow alcohol to evaporate before use.
2. Remove a TSA plate from the protective cover and carefully place in the head of the air sampler.
3. Set up the air sampler to run for 1 m^3.
4. Once sampling is complete, place lid back on agar plate. Label with location and date.
5. Repeat for the other sample areas in the unit.
6. Leave plates at room temperature for 3 days, then incubate at 32°C for 4 days.
7. After incubation, results should be read and recorded on microbiology record sheet.
8. Once plates have been read they should be disposed of by autoclaving.

3.1.4. Microbiological Monitoring of Re-circulating Cold Water Supply

Within the Cell Isolation Unit there is a water cooling system that allows refrigeration of cold blocks in the aseptic room. This avoids the need for ice, which is a potential source of contamination.

Once a month, a sample is taken from the circulating water in the water cooler to monitor the standard of the water as it is a potential source of contamination.

1. With a 1 ml sterile syringe, take a 0.5 ml sample of water from the cooler unit.
2. Transfer this sample to a TSA settle plate and allow spreading over the plate.
3. Label with sample type and date. Seal with tape.

   The procedure for incubation of the plates is as follows:

   (i) Leave plates at room temperature for 3 days and then incubate at 32°C for 4 days.

   (ii) Count the total number of any bacterial and fungal colonies present. Record this figure on the record sheet.

3.2. Microbiological Monitoring During the Isolation of Human Hepatocytes

Microbiological monitoring is carried out during isolation of hepatocytes to ensure an aseptic technique and that a clean product is produced. Hepatocyte isolation is explained in **Chapter 2** of this book.

3.2.1. Microbial Monitoring of Aseptic Technique During Processing

Settle plates are used to show the standard of the aseptic technique of an operator whilst at work in the laminar flow cabinets. Finger dabs are used to monitor potential contamination of finger tips.

1. Expose one plate on the work surface during every session in the laminar flow cabinet.
2. Place the settle plate in close proximity to the working area, but where accidental contamination of the agar surface is not likely to occur.
3. At the beginning of the session, expose the agar surface placing the lid next to the plate.
4. At the end of the session and when gloves are changed, perform finger dabs.
5. Using one settle plate, draw a line down the centre of the back of the plate. Use one-half for the right hand and the other for the left. Label right or left finger dabs.
6. Finger dabs are taken by gently touching the surface of the agar with finger tips and then the thumb.
7. Seal the plates using at least two pieces of fresh tape.
8. Label the base of the plates with the date, the batch numbers of the products produced during the session, the names of the operator(s) and the cabinet used.
9. Plates should be incubated. Leave at room temperature for 3 days. Incubate at 32°C for 4 days.
10. After incubation, results should be read and recorded on the microbiology record sheet.
11. Once plates have been read they should be disposed of by autoclaving.

3.2.2. Blood Culture Monitoring During the Isolation of Human Hepatocytes

Samples are taken at four points during processing *(2,3)* and are inoculated into a BacT/ALERT bottle, and aerobic and anaerobic bottles:

1. A sample of University of Wisconsin solution in which the liver is preserved and transported.
2. Effluent at the end of the liver perfusion step, collected at or about the time of perfusion with final buffer that contains collagenase.
3. Supernatant from cell purification centrifugation step, final wash.
4. Sample of the final product. Depending on the volume of cells isolated.

   A 50 μl of the final product is submitted for cytospin Gram stain.
5. Sample is taken by withdrawing 10 ml of solution with a 10 ml sterile syringe. Up to 5 ml per bottle minimum 100 μl per bottle.
6. Attach a needle to the syringe, leave the sheath on.
7. Wipe the bung on the BacT/ALERT bottle with an alcohol wipe and allow the alcohol to evaporate off.
8. The bottles must be labelled with a unique code to allow identification of the procedure and the stage of procedure.

BacT/ALERT cultures should be accompanied by appropriate paperwork for the institution and delivered to the clinical microbiology department.

If positive, the type of microorganism is identified to try and spot potential sources of contamination, e.g., skin flora or environmental or if they have more harmful pathogens.

Results are given on the microbiology department reports and kept with the hepatocyte isolation records. Preparations that have cultures that are positive at the final stages of the isolation are discarded or used for research.

4. Notes

1. Areas in the unit have a specified limit according to the grade of the room. As set out in Rules and Guidance for Pharmaceutical Manufacturers and Distributors: The Orange Guide *(4)*.
2. Two limits are set: a warning limit and an alarm limit. The warning limit monitors for trends and give the first indication that there might be a problem.
3. The alarm limit is the actual limit for the area. Counts the above alarm limits are recorded and the plate is sent to the hospital Microbiology Department for identification of the organism. Areas should be thoroughly cleaned with sporicidal agents. Where appropriate, other action may be taken, e.g., retrain staff.

Recommended limits for microbial contamination

Grade	Air sample cfu/m^3	Settle plates (diameter 90 mm), cfu/ 4 h(b)	Contact plates (diameter 55 mm), cfu/ plate	Glove print. 5 fingers. cfu/glove
A	< 1	< 1	< 1	< 1
B	10	5	5	5
C	100	50	25	–
D	200	100	50	–

References

1. Lehec, S., Wade, J., Mitry, R., et al. (November 2004) Evidence of microbiological screening of human hepatocytes and islets for clinical transplantation. Abstracts of 7th International Congress of the Cell Transplant Society, p. 124, Boston, USA.
2. Mitry, R. R., Hughes, R. D., Aw, M. M., et al. (2003) Human hepatocyte isolation and relationship of cell viability to early graft function. *Cell Transplant* 12, 69–74.
3. Mitry, R. R. (2008) Isolation of Human Hepatocyte. In Dhawan A. and Hughes R. D., eds. Methods in Molecular Biology: Hepatocyte Transplantation, Humana Press Inc.: New Jersey [In Press].
4. Rules and Guidance for Pharmaceutical Manufacturers and Distributors, Medicines Control Agency, 2007. Pharmaceutical Press. London.

Further Reading

Human Tissue Act 2004, Office of Public Sector Information.

A Code of Practice for Tissue Banks. Department of Health. 2001.

Guidance on the Microbiological Safety of Human Organs, Tissues and Cells used in Transplantation. Department of Health. 2000.

Alison M Beaney. (2001) Quality Assurance of Aseptic Preparation Services 3rd edition. Pharmaceutical Press.

BS 5295, Environmental Cleanliness in Enclosed Places, British Standards Institute, 1989.

Index

N

R

S

T

U

X

Y

Printed in the United States of America